Digital Agritechnology

Digital Agritechnology
Robotics and Systems for Agriculture and Livestock Production

Edited by

Dr Toby Mottram FREng, FIAgrE

Douglas Bomford Trust Chair of Farm Mechanisation,
Royal Agricultural University, Cirencester, United Kingdom

ELSEVIER

ACADEMIC PRESS

An imprint of Elsevier

Academic Press is an imprint of Elsevier
125 London Wall, London EC2Y 5AS, United Kingdom
525 B Street, Suite 1650, San Diego, CA 92101, United States
50 Hampshire Street, 5th Floor, Cambridge, MA 02139, United States
The Boulevard, Langford Lane, Kidlington, Oxford OX5 1GB, United Kingdom

Notices

Knowledge and best practice in this field are constantly changing. As new research and experience broaden our understanding, changes in research methods, professional practices, or medical treatment may become necessary.

Practitioners and researchers must always rely on their own experience and knowledge in evaluating and using any information, methods, compounds, or experiments described herein. In using such information or methods they should be mindful of their own safety and the safety of others, including parties for whom they have a professional responsibility.

To the fullest extent of the law, neither the Publisher nor the authors, contributors, or editors, assume any liability for any injury and/or damage to persons or property as a matter of products liability, negligence or otherwise, or from any use or operation of any methods, products, instructions, or ideas contained in the material herein.

ISBN: 978-0-12-817634-4

For information on all Academic Press publications visit our website at https://www.elsevier.com/books-and-journals

Publisher: Nikki Levy
Acquisitions Editor: Anna Valutkevich
Editorial Project Manager: Lindsay Lawrence
Production Project Manager: Paul Prasad Chandramohan
Cover Designer: Mark Rogers

Typeset by TNQ Technologies

Working together
to grow libraries in
developing countries

www.elsevier.com • www.bookaid.org

This book is dedicated to Dr David Llywelyn Smith, an extraordinary skilled engineer, my PhD tutor, lifelong friend and Chairman of eCow. He read the first draft of Chapter Six for me before his untimely death in 2021.

Contents

List of contributors

Simon Andrew Birrell
Department of Engineering, Cambridge University, Cambridge, United Kingdom

Ian Cox
Innovate UK, Swindon, United Kingdom

Ingrid den Uijl
Cowmanager, Harmelen, Netherlands

Narjis Hasan
VirtualVet, Kilmacthomas, United Kingdom

Josie Hughes
Department of Engineering, Cambridge University, Cambridge, United Kingdom;
Computational Robot Design & Fabrication Lab (CREATE Lab), Lausanne,
Switzerland

Youhua Li
Robotics and Autonomous Systems Group, The University of Strathclyde,
Glasgow, United Kingdom

Fumiya Lida
Department of Engineering, Cambridge University, Cambridge, United Kingdom

Patrick Joseph Lynch
RIKON Research Centre Waterford Institute of Technology, Waterford, Ireland

Paul Lynham
President's Office, The Institution of Analysts and Programmers, London, United
Kingdom

Negin Minaei
CITY Institute, York University, Toronto, ON, Canada; Urban Studies Program,
Innis College, University of Toronto, Toronto, ON, Canada

Toby Mottram
Digital Agritech Ltd, Kirkcaldy, United Kingdom

Cong Niu
Robotics and Autonomous Systems Group, The University of Strathclyde,
Glasgow, United Kingdom

Sinead Quealy
VirtualVet, Kilmacthomas, United Kingdom

Dave Ross
Agri-EPI Centre, Northern Agri-Tech Innovation Hub, Midlothian, United
Kingdom

Norbert Schlingmann
General Manager, AEF e.V., Frankfurt, Germany

Willie Thomson
Agri-EPI Centre, Northern Agri-Tech Innovation Hub, Midlothian, United Kingdom

Xiu-Tian Yan
Robotics and Autonomous Systems Group, The University of Strathclyde, Glasgow, United Kingdom

About the editor

Toby Mottram is the Founder of the eCow, Milkalyser and VirtualVet companies and an authority in the UK agricultural engineering sector. He held the inaugural position of Douglas Bomford Trust Chair of Farm Mechanisation at the Royal Agricultural University, Cirencester, UK, from 2012, where he developed a taught course in agri-technology for farm students. He began as a working herdsman and retrained as an engineer. During his career at Silsoe Research, UK, he co-invented robotic milking, cow breath sampling, in-line milk analysis and the rumen telemetry bolus. After the institute closed, he focused on the pH telemetry bolus producing a commercially viable product which has sold worldwide since 2011. The eBolus provides continuous and accurate data from within a cow's rumen to aid dairy cow management. It is supported by a web-enabled data management system. He completed a BBSRC/Royal Society of Edinburgh Enterprise Fellowship in 2016 to develop Milkalyser, a system that will revolutionise dairy cow fertility. He was elected to the Royal Academy of Engineering in 2016.

Foreword

The development of digital technologies based on powerful computing systems that can be packaged into very small physical units at relatively low cost, satellite-based and computer vision guidance systems and a wide range of sensing arrangements is now having important implications for agricultural production systems across the world. Tools based on these technologies provide the opportunity to monitor, manage and control many agricultural operations with potential benefits relating to yield, produce quality, environmental footprints, costs of production and social factors. These new opportunities provide new horizons for those concerned with developing and refining agricultural production systems and this book provides an excellent background to the available technologies as well as describing some examples of the approaches that have been used in some detail.

The pressure for agriculture to continue to provide high-quality food in sufficient quantities to feed a demanding population is likely to be an important future requirement. Such production will need to account for sustainability issues including the storage and transport from production site to end user as well as safety and welfare issues. Managing and interpreting data will be a key process that will influence the benefits that can be gained from such technologies. The authors contributing to this book have provided valuable and useful insights to many aspects relating to '*Digital Agritechnology*' and provide the reader with an excellent reference on the topic.

Professor Paul Miller
National Institute for Agricultural Botany
Cambridge, UK

Preface

I was brought up in the dairy farming area of Devon and Somerset in England and I milked my first cow aged 9, barely able to lift the teat cups onto the cow, let alone the heavy churns of milk. After milking cows for a few years and studying engineering part-time, I was lucky enough to be recruited to the UK project to develop robotic milking in 1988. Since then, I have been actively working on automation of dairy cow health sensing, and a couple of chapters in this book are the culmination of that long experience. It has been an honour to participate in the fourth agricultural revolution as we apply digital technologies to improving agricultural efficiency and thereby reduce environmental pollution. I chose the title Digital Agri-technology to distinguish this area from the chemical interventions that character-ised the 20th century. Many chemical free techniques which were abandoned in the last 100 years — hand weeding, close observation of animals, crop intercropping and rotations — can now be automated with robots and sensors replacing the boring drudgery which was the lot of the farm labourer through history.

After a long and busy career developing new agri-technologies in research, prac-tice and commercialisation and after writing many scientific papers and magazine articles, I thought when I was approached by Elsevier in 2018 that this would be a suitable and easy way to round off my career. In my naivety, I assumed that being an editor would be easy but I had not counted on developments in 2019 that led to the sale of my company Milkalyser to Lely Robotics which completed just 2 weeks before a global pandemic led to the suspension of normal life. I am indebted to the chapter authors that I recruited that they persisted with their writing when they had many other distractions. The writing process was naturally disrupted as the normal publishing cycle was also disrupted by illness and travel restrictions.

It has not been possible to include everything I would have liked, but I hope I have achieved my aim in giving a broad overview of some key technologies such as computer coding, ISOBUS, navigation and sensing for animal health. I have managed to include some chapters of discussion of wider implications of technology on how cities can be fed and how data can become marketable.

The book has also encouraged me to continue my career as I enjoy surfing this wave of technology which has barely begun to impact most agricultural systems. I had the enjoyment during the brief open period before the Omicron variant ceased travel again of visiting the FIRA conference in Toulouse where numerous autono-mous robots were on display. I had to remind the enthusiasts that robotic milking was launched 25 years ago and is still only 10% of the developed world market. This revolution has a long way to run yet and I will continue to be part of it.

I would like to thank my authors and my editor at Elsevier, Lindsay Lawrence, for her endless patience as I brought more news of delays.

An introduction to digital agritechnology

Toby Mottram
Digital Agritech Ltd, Kirkcaldy, United Kingdom

Background

For thousands of years, farming used highly anthropocentric techniques based on what humans could do from memory with power supplied by humans and with animals domesticated 10 or more millennia ago. Many farming practices were closely bound up with religious festivals and rituals that reminded congregations of the times of sowing and harvesting. Productivity and innovation were low and food trade minimal, so famines were a regular occurrence (Appleby, 1979) (Campbell & O Grada, 2011). Since the 16th century, waves of innovation have occurred, with each one strengthening and reinforcing the techniques developed in the previous period. We now have a complex global system that generates food surpluses, yet many people still go hungry due to supply chain issues rather than production. This chapter introduces key concepts in understanding the context of what is happening and describes some developments which underpin the digital revolution in agriculture.

First agricultural revolution (1700 onwards)

In 18th century England, an agricultural revolution began wherein new techniques were developed based on experimentation. Techniques such as land enclosure, field drainage, selective breeding, trailed metal cultivations replacing wooden tillage, and nitrogen supplementation (using human and animal manures) provided food for a human population that doubled in the United Kingdom from about 6 to 13 m by 1850. By then, agricultural research had become regularised with the foundation of research stations such as Rothampstead (1843) and academic centers such as Royal Agricultural College (1845). This revolution was based on Enlightenment thinking to move away from tradition to experimentation and dissemination by printed and other means. Agriculture has rarely been so important in the United Kingdom as it was in the age of improvers.[1] Even in those times, however, the most radical innovations (steam power, chemistry, electricity) were not driven by

---

[1] Northern Rural Life in the Eighteenth Century FIRST IMPROVERS—AN EARLY AGRICULTURAL ESSAY nine.

Digital Agritechnology. https://doi.org/10.1016/B978-0-12-817634-4.00005-7

agriculture. Malthus (1798) developed a hypothetical mathematical model predicting mass starvation, and since then, agricultural productivity has been disproving the hypothesis with continuously changing technologies.

Second agricultural revolution (1914—1980s)

There was a pressing need for new technology as populations grew rapidly through the 19th century. In the United Kingdom, domestic production was insufficient to sustain the population and was supplemented by expanding trade within the empire in addition to a developing global food trade. Cereals were imported after 1846 with the repeal of import tariffs (Corn Law) and refrigerated meat and dairy products after the 1870s. However, as other populations grew worldwide, competition for food resources required universal technical solutions. In the *Wheat Problem* (1917), first presented at the British Association meeting in 1898, Crookes discussed the declining yields of cereals grown on the newly cleared lands of Canada, the United States, and Australia. The initial soil stores of minerals of nitrogen, phosphorus, and potassium required supplementation, and the reserves of nitrate from guano manure, largely from the Pacific Islands, were nearly depleted. The invention of the Haber—Bosch process solved the nitrogen problem, leading to the second agricultural revolution. The increase in yield achieved by moving from the 6 Kg N/Ha deposited by atmospheric deposition to 100 Kg N/Ha resulted in huge increases in yield, although often at the expense of pollution of water and air with N compounds, with varying degrees of harm. The introduction of synthetic nitrogen fertiliser has allowed a huge increase in population over the last 100 years (see Fig. 1.1).

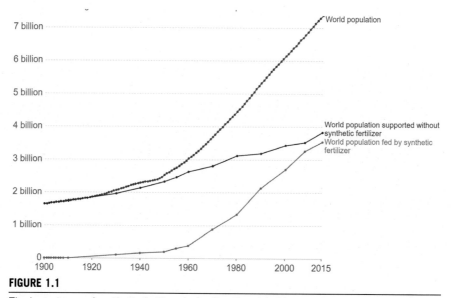

FIGURE 1.1

The importance of synthetic fertiliser in feeding the global human population in the 20th century.

Third agricultural revolution (1920s—present)

The internal combustion engine and rural electrification enabled massive improvements in labor productivity. Likewise, breakthroughs in plant breeding, herbicides, and pesticides (the Green Revolution) enabled major improvements in yields, particularly in the developing world. Human and animal motive power was replaced which has removed the drudgery, massively increased capabilities, and provided the technological platform for automated control of machinery.

The limitation of the second and third agricultural revolutions was the problem of open-loop control, where the marginal gain did not account for the costs of externalities such as environmental damage or product appeal. This led to many agritechnology techniques, particularly chemical applications, genetic manipulations, and others becoming controversial, as they can lead to pollution, environmental degradation, and consumer resistance.

In parallel, particularly in the richer parts of the world, there has been a loss of a willing and skilled rural workforce to do many jobs that are hard to robotise, such as picking vegetables and fruit. Animal welfare management also requires patience and a presence that humans find hard to consistently maintain.

Fourth agricultural revolution (1970s onwards)

In 1990, the AgEng conference series, by chance, was held in Berlin only a few months after the collapse of the Soviet system and its domination of East Germany. The conference atmosphere was very optimistic, and for the West Germans, triumphal. I remember very clearly a paper by Professor Schon discussing the revolution in agricultural productivity that we had witnessed, caused by applying power (the internal combustion engine and electricity) to machines. These machines had enabled massive increases in efficiency, allowing a single person to do the work that 50 persons would have performed during the pre-industrial age. He stated that we would now apply intelligence through computing and sensing that would have a similar effect on productivity. Sadly, he died before he could realise his predictions for a fourth industrial revolution applied to agriculture. His paper has influenced my thinking ever since, and I have been lucky enough to participate in this revolution. This is much more than making machines a bit more efficient with sensors and electronics. With technology currently in development, we can completely change our systems for supplying food to a largely urban population. Autonomous machines and systems have the potential to completely replace humans in agriculture except for supervisory and strategic planning roles, and even those may be redundant once we learn to trust command and control by computer algorithms. Like the first industrial revolution, this process may take decades and never be fully realised everywhere, given the huge need for investment and a nostalgia (particularly by politicians) for the good old days of hard physical labor and a populated countryside.

This fourth agricultural revolution is mostly driven by engineering technologies, and this textbook assumes these technologies exist in the background to provide the base for digital agritechnology. The aim of digital agritech is to use closed-loop control to improve agricultural productivity and limit environmental damage.

Basic technologies for digital agritechnology

That 1990 prediction of Professor Schon's was prescient, as then we had no internet, mobile data, solar panels, cheap digital cameras, drones, or smartphones. Computers were bulky, expensive items and had to be programmed in specialist languages. Since then, many inventions have arrived and been developed that have made an impact on digital agritechnology. This book assumes the availability of these underlying technologies as the base of the fourth agricultural revolution.

Microchips

The initial invention of transistor switches enabled the development of new electronic systems—radio, TV, etc. However, only the invention of silicon etching techniques reduced the size of transistors to nanometers and dramatically decreased the cost, enabling device size, power requirements, and costs sufficiently for agricultural use. The description of the cost reduction was characterised in a prescient description known as Moore's Law (see Fig. 1.2).

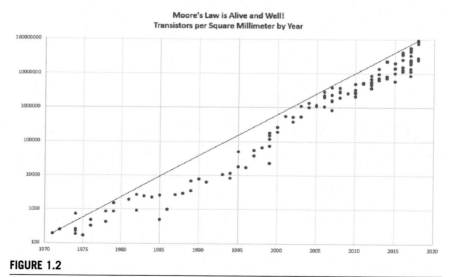

FIGURE 1.2

Silicon switch density keeps growing and has reduced the cost of computers and other devices.

Ritchie H, and Roser M, 2020, Reproduced from OurWorldInData.org under CC-YY licence.

While this relationship is most often used in the context of large-scale computing, its impact has been felt across all manner of devices from digital-image-capturing systems to microcontrollers, which are becoming the standardised nuts and bolts of digital systems. Such components are cheap enough to become ubiquitous. The reduced memory cost and increased speed have enabled other innovations, such as large-scale databases, to grow.

Electronic identification (EID or RFID)

Passive transponders for cattle identification first became available in the 1970s, but they enabled low-cost identification of animals only after becoming really cheap in the 1990s. Despite some range limitations, these devices enable new management practices for animals, such as robotic milking of dairy cows and automatic sorting of sheep flocks.

Silicon sensors

Feynman (1959) demonstrated that at the nanoscale, several physical measurements are possible. Over the years, silicon etching techniques have provided us with sensors such as accelerometers, airflow sensors, and charge-coupled devices at low prices. These have improved our ability to measure many parameters. The addition of modeling has allowed us to create proxy measures of subtle biological events such as the animal disease effects on behavior.

Global positioning systems

Global positioning systems were initially developed by the US military in the 1970s. These and the newer global navigation satellite systems allow civilian applications to locate positions around the planet within a couple of meters. Since 2000, GPS NAVSTAR encoding has been open to civilians and rival systems (Europe's Galileo, Russia's GLONASS, and China's BeiDou systems) allow accurate positioning, which is a huge benefit for field vehicles, although the encoding is not yet accurate for unaided navigation and plant finding in open fields. More of these systems are also deployed to cover specific national territories.

Meteorological information

Modern meteorology can now supply detailed weather forecasting down to 1 km resolution in developed countries and has become accurate up to 7 days ahead. This has had a major impact on the planning and execution of field operations and massively reduced weather risks even as the global climate has become more active due to global warming.

Digital imaging

The development of etched silicon transducers enabled the creation of small dense arrays of optically sensitive junctions to allow captured light to generate digital color images. These have become so cheap that they have become useful for several applications, many of which have yet to be invented. In addition, software code libraries now enable new applications to be rapidly developed to image plant details and detect animals and humans in dangerous locations.

Communications systems

Since 1990, there have been huge changes in communications. The hardware of the physical layer of the internet (as well described in *Tubes* by Andrew Blum, 2012) has developed immensely with fiber optic cable and routers for fixed locations providing the internet spine worldwide. For the final link to mobile devices, airband capability has massively expanded to provide bandwidth in the free-to-use unlicensed industrial, scientific, and medical radio bands to carry messages from digital devices used in agriculture and elsewhere. The 433 MHz band is most popular for long-range data transmission, whereas the 2.54 MHz band is used for close-range transmission of huge amounts of data, including video streaming in domestic and commercial locations. The chipsets and antennas for these frequencies have become cheap and easier to interface. The development of 5G to use higher and gap frequencies will also enhance future bandwidth. The industry prefers the term Long-Term Evolution (LTE), as there are continuous enhancements in software and antennae design.

Internet protocols

The software layer of the internet has stabilised, with the concept of packets well established. Every web page, video, image, audio, and email comes as a series of packets carrying a unique source and destination, so it can be routed through to its destination with other packets and reassembled into the full message. Networks that ship data around in small packets are called packet-switched networks. The internet relies on this technology, which is transparent as far as address and destination are concerned, although the content can be encrypted. An important subelement is the world wide web which provides an addressing system and simple protocols (GET, etc). The World Wide Web provides a Uniform Resource Locator (URL) for any location on the internet so that digital objects (documents, images, video, audio) can be linked and transferred easily.

Databases

When data was first digitised in the 1950s, complex software coding was developed to enable data sets to be efficiently linked and accessed. The conceptual structures of formal design and modeling are still used to explain the relationships. In practice, however, the speed of modern processors and memory permit inefficient structures such as files of comma-separated values to continue to dominate database input and output. Structured Query Language and its evolutions now dominate database searching. Of the systems currently in operation, almost all store huge amounts of data that can be searched and analyzed.

How this book is organised

Working knowledge of these underlying technologies is a necessary skill for all agricultural engineers, and some of this knowledge is necessary to understand the chapters of this book. Chapters 1, 2, and 3 focus on underlying digital technologies. Chapter 4 describes robotics being developed for harvesting field vegetables, while Chapter 5 describes the development of mapping and route-planning techniques for field rovers. Chapters 6 and 7 focus on sensing technologies that can be deployed to separately monitor individual dairy cows for fertility management and health management. Chapters 8 and 9 describe the potential for digital technologies to influence the resilience of urban food production systems and urban agricultural technologies close to the point of consumption. Chapter 10 provides an example of digitising the supply chain of animal products to manage data interchange and ensure that data are curated, accessible, and valued. Chapter 11 identifies some of the risks of developing digital technologies in agriculture.

Chapter 2, From data to information, is by Paul Lynham, a farmer and programmer of great experience who is now the president at the Institution of Analysts and Programmers in the United Kingdom. He focuses on how the edifice of software is built on the digital representation of data and data structures. This is an essential primer for those not experienced in software and a great revision for those who regularly work in software. It looks at how data can be secured and how data formats affect information and how information is represented and can be manipulated. This chapter focuses strongly on data integration from multiple sources, which is a key component of digital agritechnology.

Chapter 3 is about ISOBUS Standards and uses for data from farm machinery. The nature of farm machinery since the invention of the tractor has been modular and interoperable. A farmer could choose implements from different suppliers and manually control and power them from standardised tractor unit interfaces. This interoperability has been continued in the digital age by companies that manufacture most machines. They have voluntarily funded the Agricultural Industry Electronics Foundation (AEF) in Germany to standardise and test the ability of different pieces of equipment to communicate in the ISOBUS protocol. The standardisation starts at the physical layer with the design of plug connectors through to the formats of data interfaces. Chapter 3 was written by Norbert Schlingmann from the AEF, who describes the current state of ISOBUS and how new enhancements are brainstormed and agreed upon by working parties drawn from different companies.

Chapter 4, Field robotics for harvesting, is by Josie Hughes, who conducted her PhD at Cambridge and MIT and is now an assistant professor, Computational Robot Design & Fabrication Lab, at the Swiss Federal Institute of Technology Lausanne. She describes the techniques and experiments for building a robot to harvest lettuce under UK field conditions.

Chapter 5, Capturing agricultural data using AgriRover for smart farming, is led by Professor Xiu Yan of Strathclyde University. His research includes a systematic mechatronic system design methodology by developing a novel 'design process model' and associated computer support tools for mechatronic systems design. The chapter describes how a navigation algorithm can be linked to fields maps to determine the most fuel-efficient route between two points for autonomous vehicle soil sampling.

Chapter 6, Monitoring health and welfare in dairy cows, was written by Toby Mottram and Dr Ingrid den Uijl. Ingrid completed her PhD in epidemiology and biostatistics on human hemophilia in 2010, after which she returned to research in veterinary science. At the Surrey Vet School, she was involved in developing digital solutions for animal health monitoring. From 2019 until recently, she has worked for Lely Robotics as the lead in animal health monitoring. We have naturally focused on the dairy cow as the most individually highly valuable animal in farming, with plenty of exemplars of new technologies that either succeed or fail to provide accurate data.

Chapter 7, Monitoring dairy cow fertility, was written by Toby Mottram. He formerly worked as a herdsman before retraining as an engineer and being recruited to the team that developed the successful de Laval VMS robot. Since that invention, he has focused on monitoring the health of the dairy cow, particularly fertility management, and this chapter reflects the deep knowledge an engineering team must have to match the technology to the biology. He developed the Milkalyser system, which was sold to Lely Robotics in 2020.

Chapter 8, Resilient food infrastructure and location-based categorization of urban farms, was written by Dr Negin Minaei, a lecturer and visiting scholar at the Faculty of Arts and Science, University of Toronto. She was formerly at the Royal Agricultural University at Cirencester, and in this chapter, she discusses issues involved in encouraging local food production in cities.

Chapter 9, Critical review of smart agri-technology solutions for urban food growing, by Dr Negin Minaei, discusses various digital-enabled technologies to enable food production in cities.

Chapter 10, Window of opportunity for Agri 4.0: What kind of data platforms will survive? was written by Dr Pat Lynch of Waterford Institute of Technology in Ireland and Sinead Quealy and Narjis Hasan of VirtualVet. This chapter looks at the difficult issues of digitising data in the supply chain and organising a way of rewarding farmers who supply the data, the aggregators who give it meaning, and the brokers who create a market for it. Capturing and curating metadata from busy farmers is a nontrivial problem, since the human is still essential in veterinary medicine. As humans diagnose and treat animal diseases, their expertise needs to be captured in a searchable format. The data are essential to track and reduce the use of antimicrobials, which is essential for managing antibiotic resistance. They demonstrate how VirtualVet has created an embryonic platform and discuss the issues of ensuring that the value of the data is rewarded.

In Chapter 10, Risk management in digital agritech, Toby Mottram lists the apparent risks from his long experience developing digital technologies from robotic milking to inline biosensors in milking parlors.

We have tried to collect a broad picture of the wave of new digital technologies that are revolutionising agriculture; these will encourage more precision and feedback to reduce the use of chemicals and biocides, hopefully sequestering more carbon in soils and reducing pollution of the natural environment.

References

Appleby, A. B. (1979). Grain Prices and subsistence crises in England and France, 1590−1740. *The Journal of Economic History, 39*(4), 865−887. https://doi.org/10.1017/S0022050700009865X

Blum, A. (2012). *1978, Tubes, behind the scenes at the internet, pub.* Harper Collins.

CAMPBELL, Bruce M. S., & GRÁDA, Cormac Ó. (2011). Harvest shortfalls, Grain Prices, and famines in preindustrial England. *The Journal of Economic History, 71*(4), 859−886. https://doi.org/10.1017/S0022050711002178

Feynman. (1959). There's plenty of room at the bottom: An invitation to enter a new field of physics. In *Was a lecture given by physicist Richard Feynman at the annual American Physical Society meeting at Caltech on December 29, 1959.*

Malthus, T. R. (1798). *An essay on the principle of population.*

From data to information

2

Paul Lynham

President's Office, The Institution of Analysts and Programmers, London, United Kingdom

Introduction

Imagine a farmer or even a future robot, making decisions about a field operation utilising precision agriculture. Raw data is coming in from sensors on the machinery, such as crop and GPS data, whilst other data is coming from a drone, correlated against GIS information. At the same time, market and pricing data is arriving from third party cloud servers, forming a picture of the potential crop value and demand. Weather forecasts are arriving over the internet, whilst sensors measure local environmental circumstances such as current weather and soil conditions. The field operation data is also being streamed out for further analysis. Livestock positions and movements are also being tracked and stored by use of GPS or sensors located around the farm (Fig. 2.1).

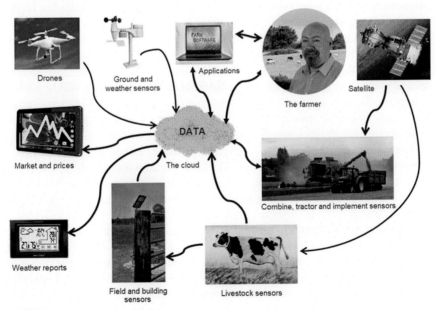

FIGURE 2.1

Farm data flows.

Digital Agritechnology. https://doi.org/10.1016/B978-0-12-817634-4.00002-1

11

Some of this data is freely available, such as GPS, prices and perhaps weather forecasts, but these can also be obtained by using commercial subscription services that may offer other benefits over their free offerings. Further data is collected on the farm, through sensors on livestock, machinery, buildings, gateways, the ground or drones. The farmer may also have some specific applications to help in the management of the farm enterprises. This information is owned by the farmer and needs to be kept secure, as much of this data could be transferred over public networks, such as the internet and stored in the cloud.

Data has value. Most people would not leave their money unprotected so it could be stolen. In a similar vein, data should be treated with care and protected, so only authorised personnel can have access to it.

This data exists in different formats, which can be processed using various algorithms, encrypted (to protect it) and transferred and stored by varying technologies and media.

This chapter will examine data, what it is and why it is important and then look at how it is represented within a computer. Data-related concepts will then be considered, such as how data can be validated, the operations which can be carried out, how it is stored, transmitted and visualised, as well as looking at issues of management and security. Finally, the current trends in how data is used and the challenges it poses will be assessed.

The future

Biometrics is the measurement and collection of data related to humans but is often used in the context of the data of living things. The use of such data in agriculture is becoming even more important than it has been in the past. Dairy cattle are usually the ones that are monitored the most amongst livestock, with many dairy farmers measuring the milk yield of their cows daily. However, with smaller low powered devices that are being developed for the **Internet of Things** (IoT), it will be possible to monitor all types of livestock on a continual basis, measuring body temperature, respiration, heart rate, movement and position, etc. By use of **Data Science**, this could give useful insights into the animal's health and well-being.

Even without implanted or worn devices, **Machine Learning** can be exploited to recognise individual animals or plants. The latter is useful for identifying weeds in a crop, so they can be removed either by mechanical or chemical means, with the ultimate goal of carrying this out without human intervention, by use of drones and robotics.

Precision Agriculture has become a key component in the third wave of modern agricultural revolution which will need each farmer to produce enough food for 265 people by 2050, up 10 times from 26 people achieved by the end of the first wave (EY Global, 2017). Precision agriculture is about timely gathering and analysing geospatial data on plant, animal or soil conditions and then setting and applying site-specific operations or treatments to optimise agricultural production and protect the environment.

With the use of drones, cameras and high-precision sensors, large amounts of data can be collected, analysed and information assimilated, allowing timely management, leading to even more efficiency gains.

The use of **Augmented Reality** to help remote professional consultants such as veterinarians and agronomists examine livestock and crops, in order to assess and carry out treatment, offers much promise, especially if combined with **Artificial Intelligence** (AI) which may be able to assist in diagnosis (Republic Lab Inc., 2018). This will involve users wearing special spectacles or goggles, which will superimpose additional graphics and metrics over what is actually being seen. It will also provide a valuable resource in training, with the ability to see inside an animal or the inner workings of farm machinery (Mileva, 2019).

Social Networks may become an important tool in connecting producers with their consumers, especially with the trends to reduce air miles, the importance of detailed traceability, food safety concerns and the drive for many restaurants to source locally produced food.

Some incredible predictions about information have been made recently (Howgego, 2019). Chiara Marletto at the University of Oxford thinks information could be the building blocks of a new theory of nature. Sophisticated manipulation of information is the thing that distinguishes living things from inert matter, so Paul Davies, a physicist at Arizona State University has proposed that information could help in a new definition of life!

Why data?

Information is our window on the universe. Information can emerge from data, which sometimes in its raw form is either unfathomable or perhaps is too large to comprehend. However, using appropriate procedures, data can be transformed so it can help in realising objectives.

William Edwards Deming, a great engineer and mathematician once said:

In God we trust, all others bring data.

Data is at the centre of any computer system. Without data, it is difficult to imagine what a computer could usefully achieve. Traditionally the following pattern is seen in computing:

$$\text{Input} \rightarrow \text{Processing} \rightarrow \text{Output}$$

This pattern has been around for a long time, even before digital computers. Charles Babbage, who is thought of as the father of computing, utilised this pattern with his mechanical analogue computers of around 200 years ago.

In the input phase, data is entered into the computer, such as being manually keyed in, or from sensors or other devices, such as another computer or storage source. An example could be the location of a cow on a farm that has been picked up from a sensor.

In the processing phase, the data is analysed or manipulated or subjected to one or more algorithms (set of instructions) in order to produce some useful information. An example could be how far the cow has moved since her last location was received. In this case, it considers the cow (by her identification data), the previous location and time together with the new location and time. From this, the distance moved by the cow and the average speed is calculated. The significance can be affected by the frequency of the inputs.

The results from the processing can now be output, such as being stored, viewed on a display or sent to another device. The results of the processing may give the consumer of this data insight into the cow's activity. Using previous data and the user's knowledge of the cow, it may lead to conclusions such as the cow hasn't moved much all day — could she be lame — or that she has been highly active — perhaps on heat?

In the above scenario, a set of raw data gathered from devices may not make much sense to a human, but a program running on the computer can process this data and can output its results with meaningful information. This information can be valuable and combined with the user's knowledge can lead to wise actions that can improve the management, health and profitability of the cow in this example.

Data in context

Data is a strange word to many people, as it is naturally used in its plural form rather than its singular (datum), but is often used as relating to one or more pieces of facts or figures. The Collins Dictionary defines it as:

Information that can be stored and used by a computer program.

On its own, data may not mean much, as it has to be put into context before it becomes information. One or more pieces of information may lead to knowledge about something and good use of this knowledge may be seen as wisdom.

DIKW model

There have been several variations of hierarchical models of data, information and knowledge; however, one that supports data through to wisdom is the DIKW model as shown in Fig. 2.2, which originated in 1994 from Nathan Shedroff who later included it as a book chapter (Shedroff, 1999).

Consider a data item with a value of 10. It does not say much without its context. At this stage, it is just data. If this is a house number that is part of a postal address, then a little more is known—it has become information. If it is associated with the city address of London and the street is 'Downing', there is now knowledge that this information relates to the residence of the UK prime minister. How this knowledge is used can be construed as wisdom.

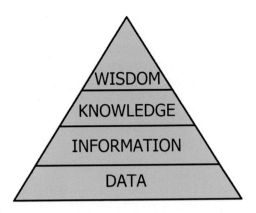

FIGURE 2.2

DIKW model.

Using **analytics**, a supermarket examined a large amount of data collected about its customers and their purchases through the use of a loyalty card. They found a correlation between the sales of nappies and beer. Further analysis found the correlation demonstrated a rise in beer sales when placed next to the nappies within the supermarket. At this point, this was just a correlation and gave the supermarket some knowledge, but what was the reason for this and how could they exploit it?

A supposition was made that young fathers went into the supermarket to get nappies for their baby. They noticed the beer and because of their fatherly commitment, had not had the time to visit the pub much, so decided to buy some beer to consume at home. By analysing the demographic data from the loyalty cards used during the purchase, the data analysts at the supermarket were able to confirm the data fitted the supposition.

With what other goods could a product placement strategy be employed, in order to increase sales for customers within this demographic? Such questions can be asked and either by using other data or perhaps by experimentation, new ways of profitably using data can be found to answer the question.

Applications

A program running on a computer consumes data and also produces it. The program uses a set of instructions (algorithms) and the data is structured so it can be meaningfully used.

An **Application Program** is software that is being used for a particular purpose (such as to aid the management of crops on a farm), as compared to **Operating System** (OS) Software, which is run to manage the computer itself. These days, App or Application is used rather than Application Program.

The way the data is structured and represented is important as that may dictate which algorithms can be used to process the data.

Data representation

With digital computers, the basic unit used for storing a value is the **Bit**. This name is made up from a contraction of the words **B**inary Dig**it**. It represents a logical value, having one of two values, usually interpreted as either **1** or **0**. Alternative interpretations could be **on** and **off** or **true** or **false**, although others are possible. Physically these values can be transmitted and stored in different ways, such as the difference in voltages in an electronic circuit, two distinct light intensities in a fibre optic or two directions of some magnetised substance.

The encoding of data in one of two values goes back to hundreds of years. Punched cards or paper tape, where a punched hole could be interpreted to do something and its absence to do nothing, were used to control patterns woven on looms. Samuel Morse used dots and dashes (Morse code) to transmit messages over telegraph.

Bits have been traditionally grouped in eights (an octet) called a **byte**. Half a byte is called a nibble. However, computers can group bits in different sized bytes, so 32-bit and 64-bit bytes are currently common sizes.

Although bits are the basic units used for representing a value, bytes or a multiple of bytes are more often used, depending on the context. Today, working memory is likely to be measured in **megabytes** (Mb) or **gigabytes** (Gb) and non-volatile storage measured in gigabytes or **terabytes** (Tb).

A **kilobyte** is often thought of as 1000 bytes, in a similar manner as a kilometre is 1000 m. However, because binary values step up in powers of 2, the nearest value to 1000 is 2 to the power of 10 (2^{10}) giving 1024. Although sometimes there is ambiguity as to whether a kilobyte is 1000 or 1024 bytes, the SI units (see Fig. 2.3) refer to multipliers of 1000. There is an ISO standard that uses different prefixes, such as a kibibyte for 1024 bytes to avoid ambiguity, but these are rarely used.

SI Units for bytes			
Decimal Value	**Power of 10**	**Prefix**	**Symbol**
1,000	10^3	kilo	k
1,000,000	10^6	mega	M
1,000,000,000	10^9	giga	G
1,000,000,000,000	10^{12}	tera	T
1,000,000,000,000,000	10^{15}	peta	P

FIGURE 2.3

SI units for multiple bytes.

Data encoding

A representation of an 8-bit byte is shown in Fig. 2.4. Each bit in the byte can represent an integer value (whole number) when it is set. Moving left to the next bit position, its value is doubled. This is in contrast to decimal, where each position is multiplied by 10 as it moves left.

Each bit can either be set or turned on, by storing a 1 or left unset, by storing a 0. The total value of the byte is attained by summing up each bit value that has been set. Looking at Fig. 2.5, if no bits are set, as in example A, then the total value is 0. However, if the first bit is set, as in B, it can be seen the total value of 1 is achieved. To get

Bit position	8	7	6	5	4	3	2	1
Value for position	128	64	32	16	8	4	2	1

FIGURE 2.4

Values at each bit position in an 8-bit byte.

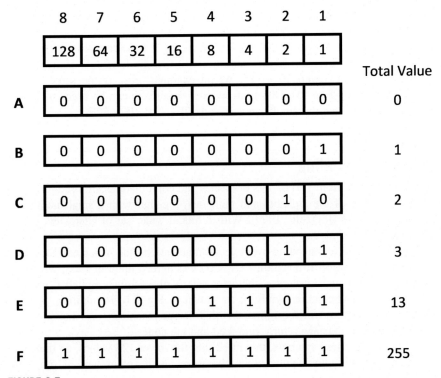

FIGURE 2.5

Example values held in a byte.

a total value of 2, as in example C, then all bits are unset except the second bit. To get a total value of 3, the first and second bits are set as shown in D, whilst to get 13, bits 1, 3 and 4 are set, i.e., values $1 + 4 + 8$. When all the bits are set, as in F, the total value is 255. Therefore, using an 8-bit byte, 2^8 or 256 different values (from 0 to 255) are available.

Using a 16-bit byte, 2^{16} or 65,536 values can be encoded, whilst with a 32-bit byte, 2^{32} or 4,294,967,296 values are possible. With a 64-bit byte, 2^{64} or 18,446,744,073,709,551,616 different values are available.

Number bases

To humans, the **binary** numbering system (using just 0 and 1) may seem unnatural, as most people are happier dealing with the decimal or denary numbering system, because of the 10 fingers and thumbs that are used to aid counting. To store decimal numbers, they are encoded in a binary format (that digital computers use), so 1101 in binary is 13 in decimal.

Binary and decimal are called number bases. A subscript can be used to avoid ambiguity when presenting values in different number bases, so 11_2 in binary is the same as 3_{10} in decimal.

Although other number bases are sometimes used in computer programming, such as **octal** (base 8) one that is worth mentioning is **hexadecimal** (hex) or base 16. Because there are only 10 different digits (0–9), the letters A–F are used to represent the values from 10 to 15. As an example, $2F_{16}$ is the same as 47_{10} ($2 \times 16 + F$ where F is 15). Different prefixes or suffixes can be used, such as **h**, **0x** or **$** to signify that the number is hex, e.g., 2Fh, $0 \times 2F$ or $2F.

Character encoding

With Morse code, from 1 to 6, dots or dashes and combinations are used to signify uppercase letters of the alphabet and numerals, together with a small number of punctuation marks (Pretzold, 1999). With 8 bits, a wider set of characters can be encoded.

Originally used for encoding over telecommunications, the **American Standard Code for Information Interchange** (ASCII) was agreed upon in 1963, which encodes both upper and lowercase character sets of the English alphabet, together with digits and punctuation symbols, as shown in Fig. 2.6. In addition, 33 non-printing control codes which originated with Teletype machines were also included. These include the carriage return, line feed and tab codes.

Even with all these symbols, this fitted into the lower 128 values, so there was still plenty of room to include other characters in the other half of the table. Tables that include these other character sets are often called the extended ASCII table.

Decimal	Hexadecimal	Binary	Octal	Char
0	0	0	0	[NULL]
1	1	1	1	[START OF HEADING]
2	2	10	2	[START OF TEXT]
3	3	11	3	[END OF TEXT]
4	4	100	4	[END OF TRANSMISSION]
5	5	101	5	[ENQUIRY]
6	6	110	6	[ACKNOWLEDGE]
7	7	111	7	[BELL]
8	8	1000	10	[BACKSPACE]
9	9	1001	11	[HORIZONTAL TAB]
10	A	1010	12	[LINE FEED]
11	B	1011	13	[VERTICAL TAB]
12	C	1100	14	[FORM FEED]
13	D	1101	15	[CARRIAGE RETURN]
14	E	1110	16	[SHIFT OUT]
15	F	1111	17	[SHIFT IN]
16	10	10000	20	[DATA LINK ESCAPE]
17	11	10001	21	[DEVICE CONTROL 1]
18	12	10010	22	[DEVICE CONTROL 2]
19	13	10011	23	[DEVICE CONTROL 3]
20	14	10100	24	[DEVICE CONTROL 4]
21	15	10101	25	[NEGATIVE ACKNOWLEDGE]
22	16	10110	26	[SYNCHRONOUS IDLE]
23	17	10111	27	[ENG OF TRANS. BLOCK]
24	18	11000	30	[CANCEL]
25	19	11001	31	[END OF MEDIUM]
26	1A	11010	32	[SUBSTITUTE]
27	1B	11011	33	[ESCAPE]
28	1C	11100	34	[FILE SEPARATOR]
29	1D	11101	35	[GROUP SEPARATOR]
30	1E	11110	36	[RECORD SEPARATOR]
31	1F	11111	37	[UNIT SEPARATOR]
32	20	100000	40	[SPACE]
33	21	100001	41	!
34	22	100010	42	"
35	23	100011	43	#
36	24	100100	44	$
37	25	100101	45	%
38	26	100110	46	&
39	27	100111	47	'
40	28	101000	50	(
41	29	101001	51	)
42	2A	101010	52	*
43	2B	101011	53	+
44	2C	101100	54	,
45	2D	101101	55	-
46	2E	101110	56	.
47	2F	101111	57	/

Decimal	Hexadecimal	Binary	Octal	Char
48	30	110000	60	0
49	31	110001	61	1
50	32	110010	62	2
51	33	110011	63	3
52	34	110100	64	4
53	35	110101	65	5
54	36	110110	66	6
55	37	110111	67	7
56	38	111000	70	8
57	39	111001	71	9
58	3A	111010	72	:
59	3B	111011	73	;
60	3C	111100	74	<
61	3D	111101	75	=
62	3E	111110	76	>
63	3F	111111	77	?
64	40	1000000	100	@
65	41	1000001	101	A
66	42	1000010	102	B
67	43	1000011	103	C
68	44	1000100	104	D
69	45	1000101	105	E
70	46	1000110	106	F
71	47	1000111	107	G
72	48	1001000	110	H
73	49	1001001	111	I
74	4A	1001010	112	J
75	4B	1001011	113	K
76	4C	1001100	114	L
77	4D	1001101	115	M
78	4E	1001110	116	N
79	4F	1001111	117	O
80	50	1010000	120	P
81	51	1010001	121	Q
82	52	1010010	122	R
83	53	1010011	123	S
84	54	1010100	124	T
85	55	1010101	125	U
86	56	1010110	126	V
87	57	1010111	127	W
88	58	1011000	130	X
89	59	1011001	131	Y
90	5A	1011010	132	Z
91	5B	1011011	133	[
92	5C	1011100	134	\
93	5D	1011101	135	]
94	5E	1011110	136	^
95	5F	1011111	137	_

Decimal	Hexadecimal	Binary	Octal	Char
96	60	1100000	140	
97	61	1100001	141	a
98	62	1100010	142	b
99	63	1100011	143	c
100	64	1100100	144	d
101	65	1100101	145	e
102	66	1100110	146	f
103	67	1100111	147	g
104	68	1101000	150	h
105	69	1101001	151	i
106	6A	1101010	152	j
107	6B	1101011	153	k
108	6C	1101100	154	l
109	6D	1101101	155	m
110	6E	1101110	156	n
111	6F	1101111	157	o
112	70	1110000	160	p
113	71	1110001	161	q
114	72	1110010	162	r
115	73	1110011	163	s
116	74	1110100	164	t
117	75	1110101	165	u
118	76	1110110	166	v
119	77	1110111	167	w
120	78	1111000	170	x
121	79	1111001	171	y
122	7A	1111010	172	z
123	7B	1111011	173	{
124	7C	1111100	174		
125	7D	1111101	175	}
126	7E	1111110	176	~
127	7F	1111111	177	[DEL]

FIGURE 2.6

ASCII table.

Data types

There are several 'types' of data that is used by a computer program. As an example, if a program reads a byte with the value 1001011, how would it know if this is supposed to represent the number 75 or the letter K in ASCII? The answer is that the program uses variables (a storage location) that can be linked to both a name and the type of data stored. Hence, when a value of the named variable is read, the program knows the type of data that has been stored and can interpret this in the correct manner.

These data types are often grouped into 2 categories, namely primitive or simple data types and composite or complex data types. Depending on the programming language used, the category that the data type falls into may change, but a simple way of viewing this is to imagine the piece of data made up of a single value, rather than being a composite of other data types. One example of this is the character and string data types. A character represents a single value, whereas a string can contain many characters, i.e., it is made up of a collection of another data type.

Primitive or simple data types

So far it has been shown how integers and characters can be encoded using bits in a byte. In this section, these and other primitive types of data will be considered.

Integer types

Integers are whole numbers. Traditionally, they are divided into bytes and other integer types, such as **int**, **short**, **long** or **word** which may or may not have a sign to indicate if they are positive or negative. They hold integer values by grouping one or more bytes together.

Real numbers

Numbers that have fractional parts are called real numbers and are commonly represented in 32-bit or 64-bit bytes, using floating-point numbers. There are number formats that use a fixed point encoding, but these are rarely used.

Floating point numbers

A common standard is the floating-point IEEE 754, which has several formats for storing real numbers. One that uses 32 bits is called the short floating-point and consists of 1 bit to store the sign, 8 bits to store the exponential part and 23 bits to store the mantissa as shown in Fig. 2.7.

The sign is signified by 0 representing a positive mantissa, whilst 1 is used for a negative value. The mantissa is stored as it would occur with 1 digit before the decimal point. The exponent is added to 127 so that both a positive and negative exponent can be signified.

Issues surrounding numbers

The handling of numbers can hold hidden problems, often ones that can easily be overlooked. Some common issues are as follows:

- Overflow and underflow
- Dividing by zero
- Not a number
- Real number comparison

Overflows and underflows If an arithmetic operation is attempted to create a value that is outside the range that it can represent, either larger than the maximum or

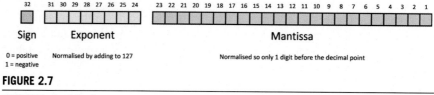

FIGURE 2.7

IEEE 754 short float.

smaller than the minimum, an overflow or underflow occurs. One solution is to set up an error handler for this situation, so it can be dealt with gracefully.

Divide by zero If an attempt is made to divide a number by 0, then the answer will be infinity. Such attempts usually cause an error, so check the value of the divisor before attempting a division.

Not a number Although the square root of a number can be negative, such as −5 and 5 for 25, if an attempt is made to calculate the square root of a negative number, then the result is not a number (NaN), i.e., this is impossible. An attempt to do this could cause an error, so checks should be made to ensure the number is always positive when trying to calculate its square root, e.g., do not attempt to find the square root of −25. There are several other scenarios in which a NaN could occur.

Real number comparison Floating point numbers cannot precisely represent all real numbers, so 2 numbers that would normally be considered equal may result as not equal when compared. They may differ by the last digit on the 24th decimal position! To prove this, if 1 is divided by 3 on a computer or calculator, the answer usually displayed is 0.3333333 reoccurring. If this is multiplied by 3, then 0.9999999 reoccurring is seen, rather than 1 which would be expected ($1/3 \times 3$). To avoid such unexpected results, the language may have a method to compare 2 real numbers which will state they are equal if there is a minuscule difference between them.

Unum

To prevent some of these issues, another number encoding system has been put forward, named the **u**niversal **num**ber, or Unum, but unfortunately few computer languages currently support it.

Its designer states that it produces more accurate answers than floating-point arithmetic yet use fewer bits in many cases, which save memory, bandwidth and power. Unlike floating point numbers, Unums make no rounding errors and cannot overflow or underflow (Gustafson, 2015). This standard may, therefore, be one for the future when take up occurs across a broad range of computer languages.

Character types

A character can be thought of as any letter of an alphabet, a digit, a mathematical or another symbol, a punctuation mark or even an invisible character that is used to control something. Examples of the latter are the tab and carriage return characters that cannot be seen but control where text flows. The term glyph is sometimes used to describe a graphical character, such as ☎ for a telephone symbol.

Historically, ASCII was a common standard used to encode characters, which managed well for languages that had a Latin script but is not good enough for other languages such as Chinese.

A character consisted of 1 byte, traditionally of 8 bits, but some programming languages offer other character types such as 'wide characters' in 16 and 32-bit forms in order to use **Unicode**.

Today, where international characters need to be used, the Unicode standard has become the de facto standard. It is used by many modern computer operating systems and supported by most computer languages.

There are 3 encodings defined by Unicode, these being UTF-8, UTF-16 and UTF-32. Unicode uses the term *code points*, where one or many code points represent a single character.

UTF-8 is backwardly compatible with ASCII, as the first 128 characters in UTF-8 correspond to the same found in ASCII. It is also widely used, with more than 94% of pages (W3 Techs, 2020) on the World Wide Web using it and the W3C recommends UTF-8 as the default encoding in XML and HTML.

UTF-16 is a variable-length encoding using 16-bit bytes, and although it is used internally in Windows, less than 0.01% of web pages on the World Wide Web use this standard (W3 Techs, 2020).

UTF-32 uses a fixed-length encoding of four 8-bit bytes, whereas the other UTF standards use variable-length encodings. This has the disadvantage of being space inefficient and is also rarely used.

Another character encoding standard is Windows-1251, which uses an 8-bit encoding design for Eastern European languages that use the Cyrillic script such as Russian, Serbian, Macedonian and Bulgarian. It is popular in Russia and is used by around 1% of web pages (W3 Techs, 2020).

Boolean

Many pieces of data only have 2 options, such as if a person is an adult or not. In this case, a check may be made to see if the person is 18 years of age or older. If the check shows the person is 18 or older, then the value *adult* will be set to **true**, otherwise, it will be set to **false**. These values are called **Boolean** values, named after a mathematician called George Boole who wrote down the laws of Boolean algebra. Usually, the values are true or false, but they can be interpreted in other ways such as yes/no or on/off.

Whilst processing data, Boolean evaluations are very common. Comparisons to see if one value is the same as another value (equality) or if it is larger (greater than) or smaller (less than) are considered evaluations that produce a Boolean result.

Boolean values can be evaluated using Boolean logical operators. The basic ones are **NOT**, **AND** and **OR**. The **NOT** operator changes the value to the opposite value. **NOT** true results in false, whilst the converse also works. If **NOT** false, the result is true. This is shown in the truth table in Fig. 2.8.

The **AND** operator is used when 2 conditions are combined. Using Boolean logic, a truth table shows the result when two variables are combined, as seen in Fig. 2.9.

By examining this table, it can be seen using the **AND** operator, both variables must be *true* for the result to be *true*.

Looking at the truth table in Fig. 2.10, it can be seen the OR operator produces a result that is true as long as one of the variables is true. However, if both variables are false, the result is false.

Operator	Value	Result
NOT	true	**false**
NOT	false	**true**

FIGURE 2.8

The NOT truth table.

Value 1	Operator	Value 2	Result
true	AND	true	true
true	AND	false	false
false	AND	true	false
false	AND	false	false

FIGURE 2.9

The AND truth table.

Value 1	Operator	Value 2	Result
true	OR	true	true
true	OR	false	true
false	OR	true	true
false	OR	false	false

FIGURE 2.10

The OR truth table.

There are other Boolean logic operators, for example, combining **NOT** and an **AND** to make a **NAND** and in a similar manner using **NOT** and **OR** to produce **NOR**.

Quite complex statements can be built up using Boolean logic, but it must be kept in mind, as with mathematics, it must be clear which operations occur first. It is better to group operations together using parenthesis, to make the order explicit.

Complex or composite data types

With the simple data types, it has been shown they have been derived solely from bytes and do not derive from another data type. With complex or composite data types, they are built from one or more of the primitive data types.

String

A string is a sequence of characters. It can be empty or it could hold a very large number of characters, but usually has an arbitrary finite length. The characters in the previous sentence could be held in a string.

If the string can change the value it holds, it is considered a variable, whilst if the string is set to a value, but cannot change later, it is called a constant. When a string appears literally in a program's source code (the computer language instructions), it is known as a string literal, e.g., 'this is a string' or 'this is another string'.

Array

Typically, an array is a collection of the same data type, stored sequentially and can be accessed using an index to signify where in the sequence the data that is being referred to exists. Commonly square brackets [] are used to denote the array.

As an example, if an array was used to store the months of the year as strings and the array is named monthsOfTheYear, then with an array starting at index 1, we would get the following values:

monthsOfTheYear[3] would contain 'March'.

monthsOfTheYear[12] would contain 'December'.

The above are examples of a one-dimensional array, but arrays can be used in several dimensions and can usually hold any data types including composite types. It is, therefore, possible to have an array of an array of strings.

Lists

Lists are also a collection of data items and in many ways share similar properties to arrays. The difference is lists are much more flexible than arrays. Often array sizes are fixed in size and it is difficult to resize them. However, lists are very flexible, so inserting a data item at a particular position in the list is easy as well as removing a data item from the list — its length can grow and reduce with ease.

Record

A record represents an entity as a collection of data 'fields'. As an example, a person record may contain fields of:

FirstName, MiddleName, FamilyName, Gender, DateOfBirth

The FirstName, MiddleName and FamilyName fields could store the person's names as *strings*, the Gender possibly as a character ('M' for Male and 'F' for Female) and DateOfBirth as a *Date* type (see later). Sometimes a record is called a *struct* which is short for structure.

Object

An object is similar to a *record* or *struct* but besides holding data about the entity or object, it also defines what operations can be carried out, usually *functions* or *methods* that can be carried out on that data.

Examples of objects could be shapes, such as a rectangle, triangle or an ellipse. The *fields* (often called *properties* or *attributes* when relating to objects) could be the height and width and the methods could be *GetArea* and *GetPerimeter* which would give the area and the length of the perimeter of each shape.

Other data types

There are some other data types, which may be available depending on the particular programming language that is used.

Other collections

Vectors, trees, stacks, queues and maps are examples of specialist collections of data items, which are orientated to keeping and accessing data in a way that will make the operations on the data efficient for a particular solution. For example, a *stack* is a data structure where data is put into a collection but the last data item added must be the first data item removed, very similar to a stack of plates (you can only take off the top plate).

Pointer

Often, the data item does not need to be passed around within a program, but instead, a pointer to it is used. This can be a more efficient way of handling data, as a pointer may be only a few bytes, whereas the piece of data itself may be many kilobytes. It is akin to a librarian having a card to represent each book in a library. If each card contains the title of the book, the author, the subject matter and its physical location in the library, then it is easier to manipulate the cards than the books themselves, e.g., if you wanted a list of all the books in alphabetical order of book title, it is far easier to arrange the cards in that order, than the books themselves.

Sub-ranges

A sub-range is a way of defining data types to limits. An example could be the months of the years represented by an integer, 1 representing January through to 12 representing December and capital letters being 'A' through to 'Z'. It allows values to be checked before being assigned to a variable of that particular data.

Enumerations

Enumerations are similar to sub-ranges, but rather than allow data values to be in a particular range, it allows the acceptable values to be explicitly listed e.g.,
cardsuit = (clubs, diamonds, hearts, spades);

Sets

Sets allow operations to be carried out on a collection of the same data types. The terminologies used for the operations are the ones found in mathematics, so 2 sets of data can be added to get the 'union', i.e., everything in the first set and the second set, or the 'difference' found between the two sets, i.e., data items that are not common.

Null

Null is a special type of data and can act as a value, a pointer and a constant. It usually represents a value of 'nothing'. As an example, if there was a collection of students and it has a method that returns a student object based upon certain criteria, such as a student's date of birth, but the date of birth for the required student does not exist in the collection, a 'null' can be returned to signify that a matching student cannot be found.

Dates and times

The format for representing a date can differ depending on where in the world it is used, as there are national standards. As an example, in the United States of America, the month is used first, then the day of the month and finally the year (MDY), whereas in most of Europe and the preferred format of the United Nations is DMY.

Because of the ambiguity of a date such as 10/12/2020 representing either December 10, 2020 or October 12, 2020 (without knowing the context), a widely used international standard of ISO 8601 is often used which is unambiguous. This uses the format YYYY-MM-DD where YYYY relates to the full 4 digit year, MM as a 2 digit month and DD as a 2 digit day. Where a month or day is a single digit, a leading zero is used. As an example, 2020-07-04 is the fourth day of July in 2020. Another benefit of this arrangement is that it is easy to sort dates in order.

Times also have formats. Using the ISO 8601 standard, the time is expressed as a 24 clock time as either hhmmss or hh:mm:ss which are hours, minutes and seconds. Because there are time zones around the earth, the time zone can also be appended. **Universal Time** (UTC), sometimes referred to as Zulu time in military circles, is a time that is based upon Greenwich Mean Time. A + can be used to signify the number of hours and minutes ahead of this time or a − for times behind. There are also time zone letters, e.g., A for 1 h ahead of UTC, B for 2 h ahead, etc.

To complicate matters further, countries or regions may use a **Daylight Saving Time** (DST), e.g., British Summer Time. This often involves advancing the clock by 1 h in spring and adjusting it back in the autumn.

Using ISO 8601, when dates and times are combined, they can be represented by concatenating (joining) the date and time together, with or without a 'T' separator between the date and the time. An example of 11:35 and 15 s on the fifth of January 2020 in London could be represented as 20200105T113515Z or using separators as 2020-01-05T11:35:15Z.

Data validation and verification

It has already been seen that by putting data in types, it is known how this will be interpreted. Using certain techniques such as using sub-ranges or enumerations, it can be ensured that only permissible values can be assigned to a particular data item. Examples of some checks are:

- Preventing the setting of a month digit to 13, when only a range from 1 for January to 12 for December is allowed.
- Ensuring a date of birth comes chronologically before a date of death for an entity.

- Checking a surname is greater than 0 characters long.

Such checks are important to help ensure the data is valid and can, therefore, be relied upon.

Some data items will have validation rules, such as a credit card number, or in the EU, a cattle ear tag number. These rules describe how a particular data item's value is constructed, such as the type of data that can be entered, the values allowed at each position and if check digits or checksums (a calculated value) are used to confirm the data is well formed and has not been changed or corrupted, either by accident or maliciously.

Therefore, valid data has undergone a quality process to ensure it is clean, correct and useful, whilst verification involves checking the accuracy and consistency of the data.

There is an old adage shortened to GIGO which stands for **Garbage In, Garbage Out**.

Even carrying out checks, errors can still occur. There are many reasons why data can be invalid or dirty. Some originate as a result of flaws in how data was collected. Human error is common and it is easy to enter incorrect data into a computer. As an example, when entering a postcode, it is easy to use an O instead of a zero or vice versa. In a similar manner, people have been known to use an I (uppercase i) for a 1, even though on paper they may look similar, to a computer they are totally different!

Besides checking the data upon entry, minimising data input from users may be a tactic and in this regard using devices and sensors that collect the data automatically may be a more reliable way of obtaining data, as well as saving time, effort and money. However, devices themselves may also become unreliable or inaccurate or may have other limitations placed upon them. It is a good policy to validate data from wherever it originates.

Sometimes it is possible to cross-check data with **reference data**, which is data that has been previously validated and verified and can be relied upon. More than one piece of reference data could be used, possibly from remote systems, to cross-check that the data that is being examined, prepared or processed is valid. A national insurance number in the United Kingdom, for example, identifies a single person, so the name, gender and date of birth can be linked to this number.

A **data dictionary** can be used to define details of all the data items of concern to an enterprise. Information about data is called **metadata** and a data dictionary can be thought of as a database of information about all the data items of concern such as name, type, constraints (such as length and format), use and description. This can then be used as a catalogue so that there is no confusion when communicating about particular data items as well as being a standard that documents their expected form and use.

Data can also come with flaws, such as being incomplete, dirty, noisy or duplicated. It can also come with a different form from that which was expected. It has been noted there are several formats for dates and this can also occur with many other types of data. In this case, the provenance of the data is important, so it is known how to translate the data to the correct form. This may be difficult or time consuming, so it is preferable to capture data in the correct form initially.

Even with all the above taken into consideration, it does not stop false data from being used. Data may conform to the correct data type, range and any other constraints, yet maybe still incorrect. Entering all the details into a system for a person named John Smith, when the data really relates to Phil Jones, may be difficult to prevent!

Data operations

The 4 basic operations of data are often shortened to CRUD, which refers to the

- **C**reation of data (making a new record)
- **R**etrieving (or **R**eading) previously saved data
- **U**pdating existing data
- **D**eleting data

Depending on the technology used to store or transfer the data, these operations are linked to keywords, verbs or concepts used in the particular technology. As an example, **SQL** is a language often used with data manipulation in databases and the keywords used in SQL mapped to CRUD would be INSERT, SELECT, UPDATE and DELETE, whereas when using **REST** to transfer data, the verbs would be POST, GET, PUT and DELETE (see Data Transmission).

Data processing

Once data has been collected, it can be manipulated to produce meaningful information. The steps taken to manipulate data can be categorised as **algorithms**. As an example, often data needs to be sorted into order, such as by date or time (chronological order) or by quantity and for this purpose there are numerous sorting algorithms. These include the **bubble sort**, **shell sort** and **quick sort**.

Other algorithms can include ones to search for items, to rank pages in a web browser, or to calculate the compound interest on a bank loan. In statistics, algorithms include ones for finding the largest value (max) or smallest value (min) and are common as well as the average (mean). To find the middle value (median), first, the data must be sorted, so in this case, 2 algorithms are used to produce the required result.

Data storage

When data is generated, it has to be stored in a form so it can be readily used. During processing in a computer, data is kept in working memory at some point, but this is usually volatile memory, i.e., the data is lost when the computer is switched off. Often there is a need to persist the data, so it can be used at a later time, even if the computer has been switched off. This is often considered to be permanent or non-volatile memory or storage.

Devices

Data can be stored in several devices which have evolved from magnetic tape drives used on mainframe computers in the past to floppy disks in previous decades, to the latest non-mechanical drives. Disc drives may be magnetic, solid-state or optical. In addition, data can be saved to data sticks (thumb or pen drives), cards, chips or SIMS (used on mobile phones) and also persisted in the 'cloud'.

The early problem with data storage on tapes was that data could not be accessed randomly — the tape had to be put at the start and spooled until the data that was required was found, i.e., sequential access. For this reason, over time, data was stored on discs, where any part of the disc's surface could be read at any time, i.e., random or direct access. This speeds up the ability to both read and write data significantly.

Modern disc storage systems use small form factor spinning discs (typically with diameters of a few centimetres) where data is stored either magnetically, often called **hard disks**, or optically, either on **Compact Disc** (CD) or one of the many types of **Digital Versatile Disc** (DVD). **Solid State Drives** (SSD) have become popular as they are non-mechanical and can be more reliable, taking up less space, besides being quicker than mechanical drives.

Data or memory sticks use 'flash memory' which are typically removable, readable, writeable, quick and light, sometimes using an integrated **Universal Serial Bus** (USB) interface. Other types are memory cards, which are often used in mobile phones or digital cameras and come in several form factors, such as **Secure Digital** (SD) cards, Mini SD or Compact Flash.

The internet of things

Asset tracking of high-value items, such as tractors or vehicles, has been around for many years. However, devices that can be used to identify an item and thereby associating it with further information have dropped dramatically in price in recent times. Thus, they can be utilised in many areas, where the intrinsic value of the item is not particularly high. Many farm supplies such as chemicals, fertilisers and feed may have a chip on the packaging or containers that can be used to identify information about the product, where and when it was manufactured and other information that is pertinent to its safe use.

Radio-Frequency Identification (RFID) chips or tags carry a unique identification number. When triggered by a nearby reader, the chip responds by transmitting its digital data which is then picked up by the reader. This is one method of **Automatic Identification and Data Capture** (AIDC). Other methods include bar codes, **Quick Response** (QR) codes and **Optical Character Recognition** (OCR).

There are application programming interfaces (APIs) available for accessing IoT information, such as **Data Distribution Service** which is a middleware protocol and standard. It is often used in real-time data exchanges between different devices, machines or systems.

The cloud and data centres

Over the last decade or so, cloud-based storage has become popular, mainly driven by mobile computing where data needs to be accessed whenever it is required and from a variety of devices, wherever they are located. It allows the availability of computer resources (both data and processing power) via the internet, without direct active management by the user.

Cloud storage can also be flexible and cost-effective, as it can be purchased as and when it is needed. This allows organisations to scale up operations when needed and to scale down later when the storage or processing is not required.

Data centres are used by organisations to share data. Traditionally, these would require hardware, including computers and network equipment to be purchased, software installed, configured and managed, even if this was needed for a relatively short period of time. However, on-demand availability is available from many large cloud providers such as Amazon, Microsoft and IBM. This solves the problem of capital investment and continual maintenance and management needed to keep a data centre operating. For this reason and many others, there has been a move away from in-house data centres, to the cloud.

There are legal regulations for different parts of the world, relating to the storage and transmission of information, such as where it is located. For example, in the **European Union** (EU) and the **European Economic Area** (EEA), the **General Data Protection Regulation** (GDPR) is laid out in EU law to provide data protection and privacy for individuals and their data. Therefore, the location of the data, be it in local servers or held in the cloud, may be important because of the legal model it is subject to for that location.

Files-based systems

File systems are used to store varied types of digital data almost as a reflection of the traditional paper-based filing systems. Files can store a single type of data, such as a text file or image file, or can hold varied data which may be in a proprietary form.

Usually, files are associated with a particular application, such as a text editor using a plain text file, a word processor (using formatted text, perhaps with embedded images) using a proprietary format, such as MS Word or Adobe **Portable Document Format** (PDF) or a spreadsheet program which save files as a workbook containing one or more worksheets.

Often the file types are denoted by the use of a file extension (a suffix attached to the file name). Typical file extensions include plain text files as txt, formatted text as doc, docx, odt or pdf, image files as png, bmp or jpeg, spreadsheets as xls, xlsx or ods and compressed files as zip or tar, etc.

To cope with a large number of files, file systems often allow files to be managed in a hierarchy, such as a drive or a volume, a folder (which may contain a series of sub-folders) and then the files themselves. The better the names of each step in the hierarchy, the easier it is to find the file that is required.

Other data about the file can be stored as file properties, such as the dates and times the files were created, the date and time they were last updated and the size of the file. Applications can display such properties in a variety of ways, using lists with columns for each property or as different types of icons, including a small picture of the contents of the file, often called thumbnails.

Databases

Where a large amount of data needs to be stored and organised, so it can be quickly accessed by one or more users, often concurrently, then individual files are not the best solution. Instead, a database can be used, which is more complex than files but allows consistent and relatively quick access of data by many and varied applications.

A **Database Management System** (DBMS) is often referred to as a **database** (DB). Historically, there were varied ways of storing data in a database, but in the 1980s, **Relational Database Systems** (RDBMS) became popular. These structure the data as a series of records, where a row is used to store the record and columns define and store individual fields of the record, very much in a similar way that data is laid out in a grid or **table** form in a spreadsheet. Records of a particular type are stored in their own database table.

It can also allow records to be linked together using a concept of **keys** and allow quick access of records by the creation of one or more **indexes** (used more often than indices). An index can be created on arbitrary information, such as a calculated or constructed identifier, or by using one or more fields of the record. This allows records to be retrieved quickly and further information brought back from other records they are linked to using their keys. This allows data to be stored and retrieved in very flexible ways.

A standard language has been devised to store and retrieve data across all relational database systems, irrelevant of the DB vendor. This is called **Structured Query Language** (SQL) and was standardised in 1986, although some vendors have added their own proprietary extensions.

SQL and modern database technology allow data to be subjected to CRUD operations consistently and safely. This may not seems obvious at first, but what happens in a stock control system that has 10 shear bolts for a baler and 2 customers who want to buy 8 each of the same bolt, from 2 different sales assistants, using the system at the same time? Databases are good at solving such problems by using various techniques, such as record locking.

In recent years, there has also been an uptake in **No SQL** databases, which stores data in a different way to the tables used in RDBMS, but has advantages in other areas. One such example is **graph databases** which give the ability to save data as a collection of nodes and edges with various relationships, allowing data to be retrieved extremely quickly. However, caution needs to be taken with such databases as their consistency is not instant and there may be a lag in data updates. Therefore, such databases are suitable for different uses to those offered by RDBMS solutions.

Databases can be spread across several computers, which may be even on different continents. This idea is used to spread the load, speeding up access and also mirror storage, providing **redundancy**, so that if one computer becomes unavailable, computers elsewhere can take up the slack. Such databases use different techniques for keeping the data synchronised, and may be costly both financially and in terms of network bandwidth, but pays premiums when disaster strikes.

Enterprise service bus

Many organisations have developed IT systems on a piecemeal basis, often in different departments, without any overarching plan or coordination. This has led to multiple applications using their own data, where in reality much of this data could be shared. This has led to several issues, such as duplicated data and not knowing which particular record holds the true state of that entity. It can also be less efficient and may demand users logging into disparate systems to enter data that is much the same for each system.

Such issues can be avoided and rectified by using an Enterprise Service Bus or **ESB**, which can be thought of as an adapter that you can plug applications into. The adapter offers common services, which can be used as required or offered to an external application as an interface. The use of **Application Programming Interfaces** (APIs) allows different applications to communicate, sharing services and data, by use of standard rules and procedures. This concept is often referred to as **Service Oriented Architecture**.

The ESB is responsible for persisting the data, so when an application or device generates data, it passes it to the ESB which decides how and where this data should be stored. This relieves individual applications from having to store and maintain their own data. When the application requires data, it asks the ESB, which is responsible for finding the data and presenting it to the requesting application. This prevents duplicate data and provides a central repository which is a **single source of truth** or **golden record**, allowing common data to be shared and utilised by different systems.

Data compression

Data can be stored in its **native form** or can be **compressed** to save space. As an example, a plain text file may contain words that occur many times. Instead of storing the word 'the' which is 3 characters, a compression algorithm could replace this with a symbol (a single character) that is not used elsewhere. It makes an index that records which symbol represents each replaced word and in doing so may cut the size of the file by a significant amount — in this example, from 3 to 1. When the file is uncompressed, the index is used to look up the symbol and replaces this with the word which was originally present when the file was compressed.

There are two types of compression, called 'lossy' and 'lossless'. The latter will never lose any data, so the data will be exactly the same when it is uncompressed. However, lossy compression will reduce the final file size by eliminating unnecessary or unimportant information. This technique is often used with large image or sound files.

There are many algorithms used for compressing data and different ones may work better with certain types of data. A common general compression format is zip (and hence the term 'zipped up' to signify compression), whilst jpeg is used with images.

Data transmission

Data has been transmitted or signalled for hundreds of years, with flag semaphore still in use today. Modern data transmission originated in telegraphy and digital data communications have evolved from the original copper wires to use of fibre optics, wireless communications and satellites. These utilise electromagnetic signalling, using varied techniques such as voltage change, radio waves, microwaves or infrared.

For physical transmission medium, fibre optic has many advantages, permitting transmission over longer distances with reduced loss of data and is immune to electromagnetic interference. For transmission without a physical conductor, wireless communication using radio waves is found in Bluetooth devices — limited to a few metres — to mobile phones which communicate across the globe.

The internet

Over many decades, several methods have been developed to transmit data between computers, such as **sockets**, which is an internal endpoint on a computer network. However, today the internet and a private version called an **intranet** are the ones commonly used for data transmission.

Originally developed for military communications, the internet has evolved into a global system of interconnected computer networks used by the public, academia, businesses and governments. It has been exploited to carry data of many kinds such as voice, email, television, radio and video streaming.

It uses a protocol called **Transmission Control Protocol/Internet Protocol (TCP/IP)** that allows data to be served up in **packets** (see Fig. 2.11), addressed and transmitted using various routes and transmission mediums, to be assembled back in order on the recipient end. It checks for data integrity (using check sums) and can request corrupted packets to be sent again, to maintain reliable and robust data transmission. It is comparable to the **Open Systems Interconnection Model (OSI)**, which breaks up the task of the transmission into layers with different responsibilities.

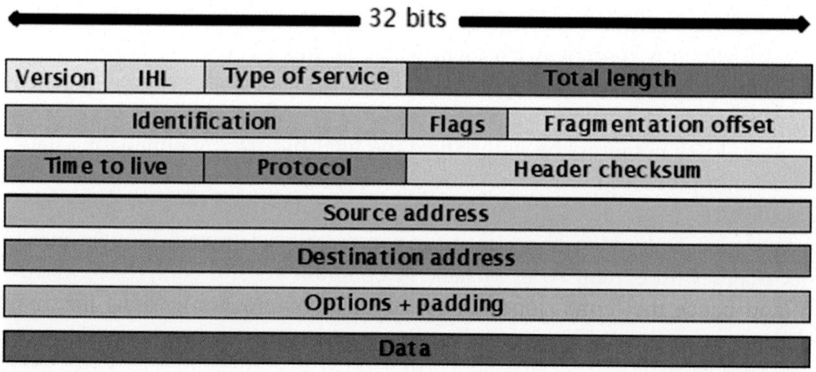

FIGURE 2.11

IP packet.

The internet has no single centralised governance and is open for malicious use by individuals or state actors. It can also be subject to surveillance and censorship to different degrees. In many parts of Europe, Africa, Canada and Australia little or no surveillance or censorship is conducted, whilst in China, Iran and North Korea, only restricted access to the internet is granted, especially where it concerns political or religious content (Wikipedia, 2020a,b).

Packet sniffer applications can be used to intercept, analyse and log data packets that are transferred over a network, such as the internet. Looking at the packet in Fig. 2.11 a lot of knowledge can be gleaned from it including where the packet came from and where it is going to as well as the data that it carries. So unless steps are taken to obscure this data, malicious users can discern a large amount of information, which preferably would like to be kept private.

The internet is by default open and insecure, so procedures need to be put in place to protect data that is transferred over it. Security software that protects against viruses, hackers and eavesdroppers and other threats should be installed on devices that are connected to the internet. Encryption can be employed, such as used by HTTPS as well as passwords, so that if data is intercepted, it is difficult to extract meaningful information.

The world wide web

The World Wide Web (often referred to as the web) is an information system that uses the internet to make resources, such as web pages, available using the **Hypertext Transfer Protocol** (HTTP). It was invented in 1989 and since then has seen phenomenal growth.

Two of its most popular abilities are to search for information, using a **Search Engine** and to display information as **Web Pages**. The latter predominantly uses a technology called **Hypertext Markup Language** (HTML) — which is often supplemented by other technologies such as **Cascading Style Sheets** (CSS) and scripting languages such as JavaScript.

Services

Historically, the sharing of data between systems would be carried out by proprietary technologies, using data formats and protocols that were closed to external systems. Today, the need for different systems to share information has driven the development of serviced based architecture (or service-oriented architecture), where systems can offer an open interface.

Using an open system does not mean that any system or user can exploit it, as in many cases, only authorised users can make use of these services. What this means is that if a user or system wishes to use the published services, they can by employing generic technologies, such as those used by the internet. This trend has led to the widespread use of APIs.

Application programming interface

A collection of procedures or functions and data structures, often produced by a third party, can be used by an application to carry out specific tasks. These traditionally were in the form of a software library or components, with the idea that it was more effective to buy in this functionality than trying to develop it in house. Examples are highly specialised mathematical functions, graphing and image processing, or any technology that is complex to exploit.

The abilities of these libraries were made available through an interface with documentation to show how the interface can be used, often with the vendors providing small example programs or pieces of code (often called snippets) to demonstrate their use. These APIs were often compiled into the program or had files that needed to reside with the program.

The modern trend is to use APIs to utilise functionality on remote systems. As an example, a patient goes to a dentist and has a radiograph (X-ray) taken of their teeth. The dentist refers the patient to a hospital, but instead of sending the radiograph in the post to the hospital consultant, an API is exploited to upload the image to a **Picture Archiving and Communication System** (PACS). When the hospital consultant sees the patient, they can look on the PACS system, with the PACS viewer utilising an API to manipulate the radiograph that was previously taken.

In many ways, this works better than having a physical radiograph present, as the consulting dentist can use the PACS viewer to examine the radiograph in flexible ways, with the ability to zoom in and out of certain areas, rotate the image around an axis and change the contrast and brightness etc. It also becomes part of the patient's record, so if the patient moves to another area and sees another dentist, the radiograph will be available if they have access to the PACS, where a physical radiograph may get lost or never get transferred to a new practice.

The use of APIs also gives the distinct advantage of being able to plug in new abilities or change parts of the system, as long as the new parts conform to the API. It is possible to build quite large and complex system by utilising APIs to join or exploit many smaller systems. This concept is often called abstraction, as the overall system is not particularly interested in the detail of how the component is implemented, but only that it conforms to the specified interface.

Data formats and messaging

Data used by the API will need to be held in a recognisable structure. In the past, a format named **Comma Separated Values** (CSV) was often used. Configuration or INI files are good at storing key-value pairs, i.e., associating a single value with a data item. However, **XML** and **JSON** formats are often better at structuring data, storing relationships and hierarchy and so are often used to hold data being transferred.

The operation of making use of a service is referred to as messaging or calling. In the past, **Simple Object Access Protocol** (SOAP) was a common way of messaging a remote system and the SOAP message body would typically carry data in XML. However, **Representational State Transfer** (REST) has become popular and this typically uses JSON to carry the payload.

REST uses HTTP (utilised by the internet) to make requests and receive responses for a remote service. If an API is designed for REST, it is called a RESTful service. As covered under the Data Operations section, REST uses HTTP verbs to carry out CRUD operations. A RESTful API makes use of a **Uniform Resource Identifier** (URI) which is a string that identifies a particular resource. As an example, if the dental application saving the radiograph to a remote PACS is taken, then the http://centralPACS.com/dentist/radiographs resource could be utilised, with the POST verb used to save a supplied image, whilst the same resource could be used to retrieve an image using the GET verb employing the supplied ID of the image. The image and the images' identifier would be sent in a JSON format. The success or otherwise of the request would be transferred back to the client application using the API as a code, known as a response. The code 404 (not found) is commonly seen by users of the web. In a similar vein, codes are used to signify success, such as 200.

Rendering

The display of information is often referred to as rendering. Data entry displays originally used character-based displays, typically consisting of 80 columns wide by 25 rows. With the advent of **Graphical User Interfaces** (GUI), often built using drag and drop visual components, more attractive screens could be crafted. User interaction became simpler and effective, as in addition to a keyboard, pointing devices, such as a mouse, tracker ball or touch screen could be utilised and more recently voice recognition can be employed.

However, many of the rich GUIs or sometimes referred to as window applications are limited to a particular platform (a combination of processor type and operating system). With the dawn of the World Wide Web, web applications became popular as they could be run from a browser without the need for software to be distributed and installed on a particular machine. On top of this, they could be run on any platform where the browser is supported, so it is possible to run the same software on Microsoft Windows, an Apple Macintosh, a Linux machine or even on mobile devices such as tablets and mobile phones, running Android or iOS.

Web-based applications can be run as a thin client model (the main logic and processing are carried out at the server side) and the browser used to display the resulting data. This fits in well with both the **Model View Controller** (a software pattern often used by web frameworks) and service-oriented architecture. To the user, it is a similar experience to viewing a static web page. However, the difference is the user can interact with it. As with web pages, HTML and associated technology are used in the browser.

Transformations

Data can be transformed to give a visual presentation. For this purpose, numerous languages can take the data as an input and produce human-readable documents as an output. Both AWK and Perl are useful for this purpose but a common solution uses XSLT and XQuery to transform XML (carrying data) to HTML (displaying data).

An example is shown in Fig. 2.12, where stylesheet templates can be used to take data held in an XML data file and transform this to HTML, to produce a document that can be viewed in a frame or a browser. In this case, the same data concerning an order can be applied to different stylesheets to either produce a delivery note or another stylesheet can be used to produce an invoice for that order. It is also possible to create a single stylesheet to produce different documents, depending on the parameters or options given.

Web scraping

Because many frameworks can be used to manipulate HTML, it is possible to utilise such frameworks to analyse web pages. Such analysis is carried out as a matter of routine by search engines that gather and monitor metrics about websites they visit. Search engine companies such as Google and Yahoo will use this data to rank and classify websites and their pages.

However, such techniques are also available to other users who may want to gather and collate information taken from web pages. This technique is often called **web scraping,** and to some, it is an important method of collecting data.

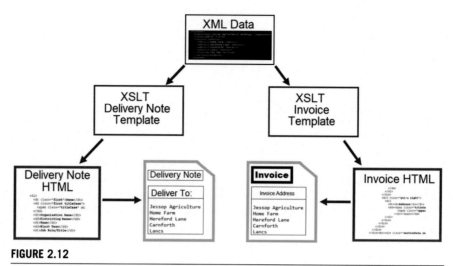

FIGURE 2.12

Data transformation.

Data visualisation

Because of the proliferation of web-based applications, there is a plethora of tools, libraries and frameworks available to enhance the look and feel of (often called **User Interface eXperience** or UIX) web-based applications. These can be used to display data in varied and novel ways, often employing graphics.

It is said a picture paints a 1000 words and often this is the case with data. Presenting data in the form of graphs or images (sometimes referred to as **infographics**) can help make the data more comprehendible, rather than presenting in textural or tabular form, as shown in Fig. 2.13.

Often graphs or infographics can summarise a data set and also allow parts of the image to be clicked upon to get further detailed information — a technique known as **drill down**. Depending on the complexity of the data, this could happen at multiple levels. Of course, the overarching goal is to allow the information to be shown and comprehended clearly, even allowing new unexpected perspectives to be revealed.

Spreadsheets are commonly used to tabulate, analyse and visualise data. Many applications such as Microsoft Excel have many built-in functions for creating tables and graphs, producing high-quality visualisations.

The use of a **dashboard** in an application to show **Key Performance Indicators** (KPIs) is becoming popular and allows users to see essential information at a glance. These can be presented either using charting, infographics or **Red Amber Green** status (RAG) with numbers or percentages.

Whichever approach is taken, good data visualisation will allow the user to view the data so it can be understood, whilst allowing them to focus on the salient points, without distorting or misrepresenting the underlying data. Good visualisation also allows a lot of data to be shown in a small space and encourage the viewer to

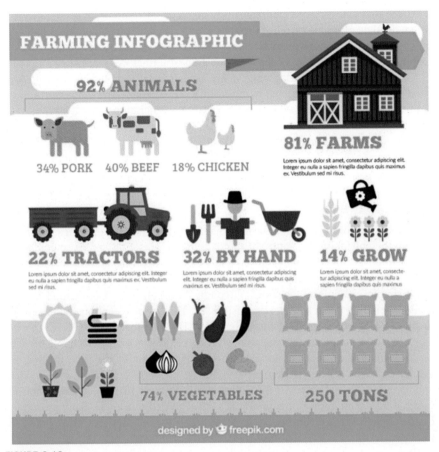

FIGURE 2.13

Farming infographic by Freepik.

compare and contrast different pieces of data. It should also allow the user to ascend the DIKW pyramid, where data is eventually transformed into wisdom.

Data classification

Data can be classified depending on who is classifying it and how the data is going to be used. For some organisations, the following categories could be useful:

1. Geographical — the location it is associated with
2. Chronological — the timeline the data is from
3. Quantitative — data that can be measured
4. Qualitative — the observed properties of the data as it cannot be measured as a number

If the first category of Geographical is taken as an example, this could look at where the data originated from at different scales, such as which continent on Earth, or which city. At a more precise scale, it could the location within a supermarket, such as an aisle or section, or the GPS location of a tractor.

However, when classifying data, many aspects need to be taken into consideration such as regulatory or legal requirements, strategic or proprietary worth, organisation goals or policies, ethical and security considerations and any contractual agreements.

The task of classifying data may form part of an **Information Lifecycle Management** process which deals with strategies for administering storage on computers.

Data management

Many organisations have their own data management processes or cycles which will involve the collection or capture of the data, its organisation and its maintenance. This also covers the data integration, analysis and reporting as well as the events that are carried out as a result of these processes.

A simple example is the **Sales Through Accurate Records** method (STAR) used by some car sales and service businesses. This involves getting detailed information on the customers and their preferences (Hinson et al., 2018). Tracking when they buy a car and then communicating with the customers on numerous events. This could be reminding them of services (using calculations of the time interval and annual mileage) (Gates & Hemingway, 2000), keeping their records up to date, enticing with special offers or evaluating when they are likely to change their car (again either from known time preferences or recorded mileages). Such predictions can be applied to the enterprise so expected ordering and sales can be predicted and managed.

In essence, the process can be boiled down to 3 stages in the cycle (EDM, 2020) as shown in Fig. 2.14. These are as follows:

- Collection — capturing new data
- Active Management — reviewing and updating the data to ensure its accuracy
- Querying or Reporting — using the data

FIGURE 2.14

Typical data management cycle.

These processes involve a myriad of other data related fields, such as **Data Capture**, **Data Analysis** and **Business Intelligence**. They will dictate the policies that will be used, e.g., for collecting data, the 5 W's (Workfront, 2018) may be used (who, what, when, where, why) to determine the kind of data that is needed. These questions apply regardless of the type of data being collected, whether it comes from a sensor, a database or from manual input and the entity it relates to.

Security

Security is a wide field, with many aspects, but it is an area that needs to be properly tackled as the consequences of not securing data can be catastrophic. Data is a precious commodity and needs to be safely and secured stored, handled in the correct manner and strategies put in place for disaster recovery. When it is no longer required, data needs to be properly erased so that it can never be accessed.

Attackers

There are different types of attackers or **actors** — people or organisations that are interested in obtaining and then exploiting data. These can be governments, often called 'state actors' and 'non-state actors' such as criminal gangs and individuals. They can attack in many ways, so threats to a system need to be scrutinised to assess any potential week points in order to take preventive measures.

Threat modelling

A logical way to examine potential security issues is to use a technique named **Threat Modelling**. This looks at the current system, its absence of appropriate safeguards, the type of attackers that may exist and how they would go about attacking the system (**attack vectors**). One such approach developed by Microsoft is STRIDE:

- Spoofing
- Tampering
- Repudiation
- Information disclosure
- Denial of service
- Elevation of privilege

Each of these threats is examined in detail, ideally by various stakeholders, who can bring their own perspectives to the table. This should initially be carried out in the design phase of a system but can be applied retrospectively if required.

CIA triad

There have been several models put forward to describe the principles of security. Of these, the CIA Triad (Commissum, 2018) is likely to be the best known (not to be confused with the Central Intelligence Agency). This lays out 3 tenets:

- Confidentiality — limit who and what can access the data
- Integrity — ensuring the data is correct and consistent and has not been tampered with
- Availability — the data can be accessed whenever it is needed

Confidentiality can be breached when unauthorised users gain access to information. This can range from someone inside the organisation accessing data they should not, external parties eavesdropping on data transmissions and others trying to unencrypt protected data.

A technique that uses a key or code to change the data into an unrecognisable form in such a way that only authorised users can access it is called **encryption**. It commonly uses a password or some other key known to the authorised user. Because of the algorithms used, it is extremely difficult to unencrypt the data and the usual attack comes from malicious parties trying to gain knowledge of or guess the password. Thus, it is a good idea to use passwords that are not easily guessed.

Integrity can be compromised by human error or accidental physical compromisation of media where data is stored, e.g., scratching or breaking a data DVD. It can also be done by maliciousness, such as hackers altering stored data.

There are many methods of ensuring the integrity of data, many of which may be supplied by ensuring confidentiality and availability. A tried and trusted technique is the use of checksums. However, **Blockchain** technologies may hold some promise in this area, as they are resistant to the modification of data. This technology is used with digital currencies, such as Bitcoin.

Availability can be denied when a disaster such as a flood or fire destroys the data or can be done maliciously using a Denial of Service (**DoS**) attack, e.g., making a large number of requests to a service so that it cannot cope and continually crashes.

Storing data in more than one place, or creating backup copies, should be part of normal data operations. These can then be used if the working copy of the data is destroyed or becomes inaccessible, by providing an opportunity for the data to be restored.

In recent years, several viruses have appeared which can encrypt files on infected systems, without the user knowing, hence making the data unavailable. The virus writer may then contact the user of the infected system and demand money to unencrypt the files. It is therefore important that computer systems have up-to-date security and antivirus software running on their systems.

A variation of the CIA principles is the addition of a further 3 attributes, named the Parkerian Hexad (Wikipedia, 2020a,b):

- Authenticity — the accuracy of the claim of ownership of the data
- Possession or control — loss of data even if a breach of confidentiality has not occurred
- Utility — the usefulness of the data

Regulation

The **General Data Protection Regulation** (GDPR) regulation came into effect within the EU in 2018 (Intersoft Consulting, 2016). This lays down standards for personal data handling and the procedures for data breaches. Violators of the GDPR can be liable for up to 20 million euros or 4% of their annual worldwide revenue, whichever is the greater. It is therefore paramount that data is securely protected.

One tactic is to use anonymisation or pseudonymisation to protect data. The former allows data to be used but does not associate it with a data subject, so irreversibly destroys any way of identifying the data subject. Pseudonymisation substitutes the identity of the data subject in such a way that additional information is required to re-identify the data subject.

Politics

Data and its security is topical and often subject to politics, be it the police wanting access to data stored on a suspect's mobile phone (Nicas, 2018), to President Donald Trump signing new laws in the United States banning TikTok and WeChat (BBC News, 2020), which are two of China's biggest apps over perceived security threats. When the Taliban took control of Afghanistan, many people there tried to delete their 'digital footprint' which Western Powers had set up but not fully secured (Dad, 2021).

The law will always be well behind what is done with data and until it catches up, there will always be actors who will take advantage of a situation. Even so, there have been several incidents with the 'Track and Trace' initiative in the United Kingdom (Cellan-Jones, 2020), with the Department of Health conceding the initiative to trace contacts of people infected with Covid-19 was launched without carrying out an assessment of its impact on privacy. Data collected in pubs and restaurants for tracing purposes has also been misused (White, 2020), with reports of bar staff illegally texting female customers for dates.

Data trends

With increasing abilities of both processing speeds and storage capacities, the ability to manipulate large amounts of data has opened up many possibilities. Adding value to data is important and many tools are available to help in this regard.

Data Science has become an important field in information technology, using various methods, processes and algorithms to gain knowledge from data. The analysis of an organisation's data has the potential to reveal insights into its operations and how it can improve its performance or serve its customers or users better, leading to both efficiency and financial gains. Associated terms are **Data Warehousing** for storing and reporting on data, **Data Mining** for discovering patterns in large data sets and **Data Analysis** for cleaning, transforming and modelling data to discover useful information.

Big Data is a term coined to denote that some data sets are too large or too complex to be treated traditionally. It utilises different techniques such as observing patterns rather than analysing individual data items. There can also be problems with large amounts of unstructured data, which may be unsuitable for storage and analysis in a traditional RDBMS because it could need a considerable amount of pre-processing before being useful.

Big Data has a myriad of uses. As an example, digital agriculture and big data will change the way seed and agrichemical companies market, price and sell their products as well as how they select and invest in their research and development pipeline. It will guide how they recommend and technically support product sales, manufacture and distribute products and manage credit and financial risk (EY Global, 2017).

Artificial Intelligence (AI) holds some promise for the future, with gains being found in focussed fields, rather than General AI. One example concerns an application produced by Google Health which was trained on 91,000 mammograms taken from women in the United Kingdom and United States. It has proven to be better than a radiologist at spotting signs of cancer (Hamzelou, 2020).

Founded by Sirs Tim Berners-Lee (inventor of the World Wide Web) and Nigel Shadbolt, the **Open Data Institute (ODI, 2020)** is a non-profit organisation with a goal to connect, equip and inspire people around the world to innovate with data. It has advocated open data as a public good, emphasising the need for effective governance models to protect it.

In 2020 around 50 zettabytes (10^{21} bytes) will be processed but this will at least triple by 2025, with currently around 6% being used on data processing (Gent, 2020). Some predictions point to 10% of the world's electricity being used for this purpose by 2030. This is unsustainable, so more efficient processing is required, including hardware with lower power consumption as well as better algorithms and techniques for data processing.

Challenges

Data is already important to agriculture, but there will be even more reliance upon it in the future — it will become critical to any farming venture. Once it is collected and processed, it needs to be understood and interpreted, so it can provide guidance to current and future management and strategy.

Volumes of data collected will increase radically and the majority of this is likely to be stored in the cloud, but machine learning will augment many processes (Itransition, 2020), so that human intervention will be reduced or even eliminated.

Unstructured verbal data is a real challenge that is being poorly addressed by current systems. VirtualVet (2020) is a company trying to solve that problem. They aim to enable the digitisation of important animal treatment data early and once. By use of smartphone devices, they bring significantly stronger standards of animal health traceability to the food-producing value chain. In addition, the digitised data has value to various stakeholders in the industry.

Natural language processing is certainly one area where machine learning could have a drastic impact. Such systems could likely detect and predict new disease patterns, but there are ethical issues involved. In a similar vein to consumers wanting to know where their food comes from and how it is produced, people will also want to know how and on what basis the information is being presented, e.g., how did the technology arrive at its results? It sounds simple, but explaining how this is accomplished is challenging.

Data science is relatively new, but the demand for data scientists is high and as volumes of data grow, the demand for these specialists will increase. How this demand is met is yet another problem.

Security and privacy is an ever-escalating issue, especially with growing rates of data and increasing levels of cyber-attacks. Data protection will need to keep up with the game-changing conditions. There is a lack of skills in this area, caused by a lack of education and training and there is a prediction that there will be over 3.5 million cybersecurity job openings by 2021 (Cybercrime Magazine, 2017) but a 1.8 million worker shortfall by 2022 (Kawamoto, 2017).

However, most authorities agree that big data will mean big value, bringing more job opportunities for the right people. Many enterprises will shift from being data-generating to data-powered, making use of actionable data and business insights.

References

Intersoft Consulting. (2016). *General data protection regulation* [Online] Available at: https://gdpr-info.eu/. (Accessed 9 March 2020).

Itransition. (2020). *The future of big data: 5 predictions from experts for 2020-2025* [Online] Available at: https://www.itransition.com/blog/the-future-of-big-data. (Accessed 27 July 2020).

BBC News. (2020). *TikTok: President Trump signs orders to ban it in the US within 45 days* [Online] Available at: https://www.bbc.co.uk/newsround/53620689. (Accessed 8 October 2020).

Cellan-Jones, R. (2020). *Coronavirus: England's test and trace programme 'breaks GDPR data law'* [Online] Available at: https://www.bbc.co.uk/news/technology-53466471. (Accessed 8 October 2020).

Commissum. (2018). *The CIA triad: The key to improving your information security* [Online] Available at: https://commissum.com/blog-articles/the-cia-triad-the-key-to-improving-your-information-security. (Accessed 9 March 2020).

Cybercrime Magazine. (2017). *Cybersecurity jobs report vs survey* [Online] Available at: https://cybersecurityventures.com/jobs-report-vs-survey/. (Accessed 27 July 2020).

Dad, N. (2021). Data privacy in a war zone. *New Scientist, 251*(3349), 23, 28 August 2021.

EDM. (2020). *The data management cycle* [Online] Available at: https://effectivedatabase.com/the-data-management-cycle/. (Accessed 9 March 2020).

EY Global. (2017). *How digital agriculture and big data will help to feed a growing world* [Online] Available at: https://www.ey.com/en_gl/advisory/how-digital-agriculture-and-big-data-will-help-to-feed-a-growing-world. (Accessed 8 March 2020).

Gates, B., & Hemingway, C. (2000). Know your numbers. In *Business @ the speed of thought: Succeeding in the digital economy* (p. 224). London: Penguin.

Gent, E. (2020). Number crunch. *New Scientist*, 40, 14 March.

Gustafson, J. L. (2015). *The end of error: Unum computing* (1st ed.). Boca Raton, FL: CRC Press.

Hamzelou, J. (2020). AI system is better than human doctors at predicting breast cancer. *New Scientist*, 1 January https://www.newscientist.com/article/2228752-ai-system-is-better-than-human-doctors-at-predicting-breast-cancer/, 2020. (Accessed 31 March 2022).

Hinson, R. E., Adeola, O., & Amartey, A. (2018). Twenty four attributes of outstanding sales-people. In *Sales management: A primer for frontier markets. s.l.* (p. 20). Information Age Publishing.

Howgego, J. (2019). How to think about information. *New Scientist, 14*, 37. December 2019 (Special Issue).

Kawamoto, D. (2017). *DARKReading* [Online] Available at: https://www.darkreading.com/careers-and-people/cybersecurity-faces-1-8-million-worker-shortfall-by-2022. (Accessed 28 August 2021).

Mileva, G. (2019). *How augmented reality could revolutionize farming* [Online] Available at: https://arpost.co/2019/01/18/how-augmented-reality-could-revolutionize-farming/. (Accessed 8 March 2020).

Nicas, J. (2018). *Apple to close iPhone security hole that law enforcement uses to crack devices* [Online] Available at: https://www.nytimes.com/2018/06/13/technology/apple-iphone-police.html. (Accessed 8 October 2020).

ODI. (2020). *About the ODI* [Online] Available at: https://theodi.org/about-the-odi/. (Accessed 13 January 2020).

Pretzold, C. (1999). In *The hidden language of computer hardware and software* (p. 9). Redmond: Microsoft Press.

Republic Lab Inc. (2018). *How augmented reality can bring changes in agriculture?* [Online] Available at: https://www.republiclab.com/augmented-reality-agriculture/. (Accessed 8 March 2020).

Shedroff, N. (1999). Information interaction design: A unified field of theory of design. In R. Jackson (Ed.), *Information design* (p. 267). Cambridge: MIT Press.

VirtualVet. (2020). *We're the health in a healthy food supply chain* [Online] Available at: https://www.virtualvet.eu/. (Accessed 27 July 2020).

W3 Techs. (2020). *Usage of character encodings broken down by ranking* [Online] Available at: https://w3techs.com/technologies/cross/character_encoding/ranking. (Accessed 12 January 2020).

White, N. (2020). *Creepy bartender uses coronavirus contact tracing data to ask out a girl he gave a free drink to - as Australians are warned their personal information could be misused or stolen* [Online] Available at: https://www.dailymail.co.uk/news/article-8516533/Creepy-bartender-uses-coronavirus-contact-tracing-data-ask-girl.html. (Accessed 8 October 2020).

Wikipedia. (2020a). *Internet* [Online] Available at: https://en.wikipedia.org/wiki/Internet. (Accessed 28 June 2020).

Wikipedia. (2020b). *Parkerian Hexad* [Online] Available at: https://en.wikipedia.org/wiki/Parkerian_Hexad. (Accessed 9 March 2020).

Workfront. (2018). *Project Management 101: The 5 Ws (and 1 H) that should be asked of every project!* [Online] Available at: https://www.workfront.com/blog/project-management-101-the-5-ws-and-1-h-that-should-be-asked-of-every-project. (Accessed 9 March 2020).

ISOBUS — standards and uses for data from farm machinery

3

Norbert Schlingmann

General Manager, AEF e.V., Frankfurt, Germany

Introduction

This chapter describes the ISOBUS and what is needed to get cross-manufacturer compatibility of electric and electronic components. Together with over 250 international member companies, the Agricultural Industry Electronics Foundation (AEF) has been committed to cross-manufacturer compatibility of ISOBUS products since 2008. The aim is to ensure smooth interaction of machines, devices and terminals in agricultural practice.

Although manufacturers of agricultural technology have agreed on ISOBUS as a common "language" for agricultural applications on a worldwide basis, not all machines, devices and terminals are able to communicate with each other for various reasons. ISOBUS is based on the International Standard ISO 11783. In practice, the standard is large and open for interpretation causing some confusion and dissatisfaction amongst dealers and farmers.

To enhance quality and transparency, the AEF has drawn up supplementary guidelines for individual functionalities. During development, AEF members design products according to the functionality concept. Subsequently, they perform a conformance test, also developed by the AEF, to have their ISOBUS products certified by independent test institutes. If a product has successfully passed the test, the manufacturer may use the AEF ISOBUS certified label as evidence. It documents the conformity of the ISOBUS product with the ISO standard 11783 and the AEF guidelines. Additionally, the certified product is published in the AEF ISOBUS Database. This database — which is also available as an APP — quickly provides detailed information to dealers and farmers about the specific ISOBUS functions of a product and the compatibility with other products.

Short history of ISOBUS

ISOBUS is a serial control and communication data network based on the ISO 11783 standard, for agricultural and forestry equipment. It's a communication protocol for the agriculture industry based on CAN bus technology. Electronics are the

Digital Agritechnology. https://doi.org/10.1016/B978-0-12-817634-4.00001-X

key to making machinery more efficient, precise and economical. ISOBUS is one of the most important techniques for this industry. The idea behind ISOBUS is 'plug and play' with any tractor-terminal-implement-combination. ISOBUS also includes the data transfer between these mobile machines and farm software applications. It is the most significant and comprehensive standard to date, but it leaves room for interpretation, which has led to a great number of innovative, but proprietary, ISOBUS solutions.

The introduction of ISOBUS products into the market in the mid-2000s did not go as intended. Equipment manufacturers were taking different approaches in engineering interpretation since the ISOBUS standard was really complex and written from a theoretical point of view in the ISO working groups. Also naming of ISOBUS products and marketing information were confusing to end-customers and dealers.

How well the components of an ISOBUS system work together depends on what they have in common, i.e., the functionalities supported by all of them.

In order to minimise the room for interpretation, the Agricultural Industry Electronics Foundation, AEF, was founded in 2008. It develops guidelines which add to the ISO standard and thus allow for precise definitions of ISOBUS functionalities. Ultimately, this provides for clarity regarding the compatibility of ISOBUS products.

The uniform ISOBUS language enables standardised communication between the different cross-manufacturer machinery and brings multiple benefits. One terminal can now be used for several machines. This means 'plug-and-play' is guaranteed to any ISOBUS tractor. Using its control and operating settings, the terminal brings the implement automatically into view.

Other benefits include standardisation of the control settings, providing a better overview in the cabin, along with a simpler connection between tractors and implements, not to mention the cost savings when using several ISOBUS implements. ISOBUS communication is especially suitable for farmers who sow, spray or spread, or for machines that provide the owner with a great deal of management information.

ISO 11783 based on CAN bus

The use of CAN in agricultural systems started in the early to mid-90s, typically as a proprietary solution. Work to develop the ISO 11783 standard began at about the same time, leveraging work from the SAE J1939[3] standard and extending it in unique ways for the agricultural industry. It took several more years of development before successful interoperability was realised between various manufacturers equipment (Fig. 3.1).

Fundamental to the original design of ISOBUS, and with the size of the equipment — tractors and connected implements, is the definition of the physical layer (ISO 11783-2) which specifies the maximum length of a CAN backbone (e.g., the primary CAN harness extending from the front-most ECU to the rear-most ECU) as 40 m and a maximum bit-rate of 250 kb/s, or approximately 1800 messages per second.

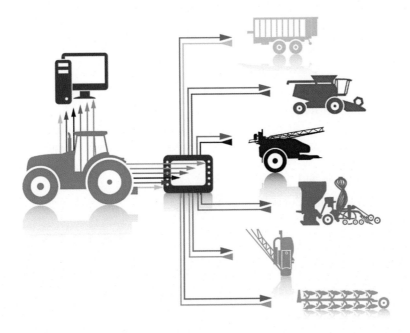

FIGURE 3.1

ISOBUS communication between various manufacturers equipment.

https://www.aef-online.org/.

In the years since that original physical layer definition, the foundational layers of the ISOBUS communication protocol were developed (ISO 11783-3, ISO 11783-5), permitting successful network management, device discovery, and communications. At this point, it had value, but there was a lot more to come as ISOBUS was extended for the specific needs of the agricultural industry.

Various application-level functionalities were developed that bring the true value of ISOBUS to customer equipment. Applications include Virtual/Universal Terminal (ISO11783-6) [2004], Tractor ECU (ISO11783-9), Task Controller (ISO11783-10) [2009], Diagnostics (ISO11783-12) [2009], File Server (ISO11783-13) [2007] and most recently, Sequence Control (ISO11783-14) [2013].

All these services were being developed and continue to be developed and mature.

Components and network

A modern ISOBUS system consists of various components, including the tractor, terminal and implement. It is always a question of the performance of the terminal and implement — and eventually, of the installed options. For increased transparency, functionalities have been defined.

ISOBUS system overview

Fig. 3.2 shows a typical ISOBUS system with the main components located on tractor and implement connected via ISOBUS breakaway connector IBBC (Fig. 3.3).

Components of an ISOBUS system:

- Tractor Electronic Control Unit (TECU)
- Universal Terminal (UT)
- Implement Electronic Control Unit (ECU)
- Joystick
- Standardised hardware
  - Connectors
  - Cables
- Task Controller

Connectors

The ISO 11783-2 standard specifies three types of connectors:

Implement bus breakaway connector (Fig. 3.3)

- The connectors are designed to be extremely robust, protected against contamination by an optional cover, and designed in such a way that no damage occurs if they are pulled off;
- The power supply to the attached implement is also provided via connections in the connector up to 60 A.

Diagnostic connector (Fig. 3.4)

- The SAE J1939/13 Standard defines a standard connector for diagnostic purpose;
- Used for easy access to the CAN bus for diagnostic purposes;
- The connector is a Deutsch HD10 - 9—1939 (9 pins, round connector).

In-cab connector (Fig. 3.5)

- The connector is installed in the tractor cab;
- In the connection cable of the terminal, only the connection to the tractor socket has been standardised so far;
- The terminal-side interface is manufacturer-specific.

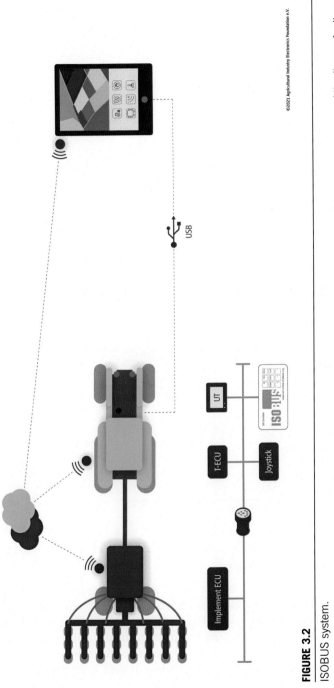

USB

©2021 Agricultural Industry Electronics Foundation e.V.

https://www.aef-online.org/.

FIGURE 3.2

ISOBUS system.

FIGURE 3.3

Implement bus breakaway connector (IBBC).

https://www.aef-online.org/.

FIGURE 3.4

SAE J1939/13 off-board diagnostic connector.

https://www.aef-online.org/.

ISOBUS functionalities

In order to solve the complexity of the ISOBUS standard, the AEF project teams defined so-called functionalities that encapsulate the different control functions in a network, such as the Universal Terminal, the Tractor ECU, an auxiliary device or a Task Controller. For increased transparency, functionalities have been defined. And by splitting up the standard into well-predefined functions, it is easier to explain to the end-user what it means when a device is said to be ISOBUS compatible.

FIGURE 3.5

In-cab connector.

https://www.aef-online.org/.

It does not necessarily mean it supports all functions, but by using the AEF Guidelines and Functionalities, a manufacturer can now clearly implement compatibility to other devices according to these specific functions.

An ISOBUS functionality is a kind of product which can be explained to the end-user as a separate 'module' on the ISOBUS. One or more functionalities can be bundled together into a retail product intended to interconnect with other products that contain AEF functionalities. In an ISOBUS system, only the least common denominator of functionalities can be used. Only functionalities supported by all components involved are available. And only then the famous 'plug and play' will work. The description of the functionalities provided here is of course highly condensed.

Universal Terminal (UT)

The Universal Terminal functionality has capability of operating an implement with any terminal; also, the ability to use one terminal for operating different implements.

Auxiliary Controls (AUX-O and AUX-N)

The Auxiliary Controls functionality is used to connect additional control elements that facilitate the operation of complex equipment, such as a joystick. There are an 'old' and a 'new' auxiliary control which are not compatible. Implements and functions certified according to AUX-N cannot be operated with input devices certified according to AUX-O and vice versa.

Task Controller (TC)

The Task Controller functionality distinguishes between three different functions: Basic, geo-based and Section Control (Fig. 3.6).

Task Controller basic (TC-BAS) describes the documentation of total values that are relevant for the work performed. The implement provides the values. For the exchange of data between farm management system and Task Controller, the ISO-XML data format is used. Jobs can easily be imported to the Task Controller and/or the finished documentation can be exported later.

Task Controller geo-based (TC-GEO) is an additional capability of acquiring location-based data — or planning of location-based jobs, for example, by means of application maps.

The Task Controller Section-Control (TC-SC) functionality controls automatic activation of sections on the implement, as with a plant protection sprayer, seed drill or fertiliser spreader, based on GPS position and desired degree of overlap.

Tractor ECU (TECU)

The tractor ECU is the tractors 'job calculator'. This provides information, such as speed, power take-off RPM, etc., for the other ISOBUS participants. For the

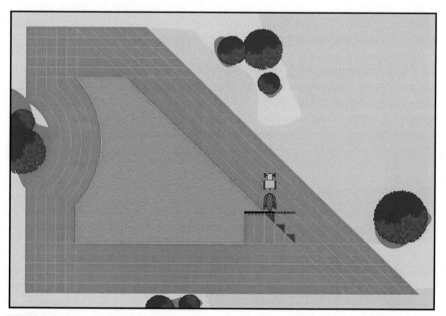

FIGURE 3.6

Task controller — Section Control: automatic switching off sections.

https://www.aef-online.org/.

certification of this functionality, a connector on the back of the tractor and a terminal outlet in the cab are needed.

Tractor Implement Management (TIM)

Whilst the communication with TECU is uni-directional, i.e., the tractor provides certain information, the TIM functionality has the capability of bi-directional communication. The Tractor Implement Management System (TIM) allows an implement to automatically control specific tractor functions, such as the forward speed or the remote valves of a tractor. By letting the implement optimise its operation, the overall system can achieve higher levels of productivity with less operator fatigue (Fig. 3.7).

Up until now, in most ISOBUS solutions, the tractor controls the implement; however, TIM is based on bi-directional communication, in other words, allowing the implement to play the leading role with ISOBUS as the basis for this communication.

The implement is only able to play the leading role if the data transmission is secure. For this purpose, the AEF has developed an infrastructure that enables secure communication based on proven standards, such as electronic banking. This standardised solution — in conjunction with digital certificates — is necessary so that the implement can control certain tractor functions and can actively carry out the work process without driver input. Both tractor and implement trust each other so that the farmer can work more precisely, effectively and economically whilst simultaneously increasing quality. TIM saves time and money. The field operation becomes technically simpler and easier as TIM takes over tiring and repetitive tasks and the farmer can use TIM with his tractors and implements from different manufacturers. TIM makes optimum use of the installed performance potential for higher productivity of the overall system. Automatic processes replace manual operation and prevent the driver trying to exceed the system capabilities.

ISOBUS Shortcut button (ISB)

The ISOBUS Shortcut button (ISB in short) makes it possible to deactivate functions of an implement that were activated by means of an ISOBUS terminal. This is necessary when the implement in question is not currently in the foreground, for example, when several implements are being controlled by a single ISOBUS terminal. Which functions an ISB is able to deactivate on an implement can vary widely and must be defined by the respective manufacturer.

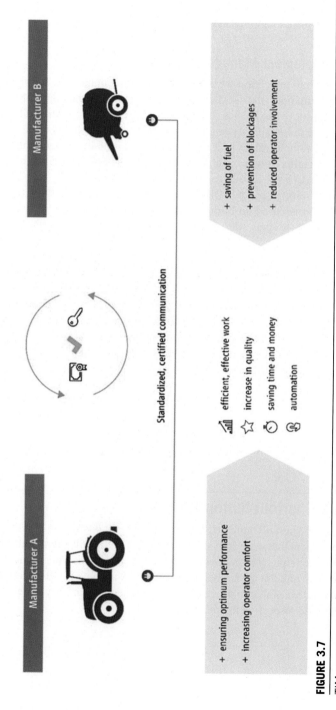

Future-proof communication between tractor and implement
TIM process control using the example of a round baler

Manufacturer A

Manufacturer B

Standardized, certified communication

+ ensuring optimum performance
+ increasing operator comfort

+ saving of fuel
+ prevention of blockages
+ reduced operator involvement

efficient, effective work
increase in quality
saving time and money
automation

https://www.aef-online.org/.

FIGURE 3.7

TIM process control using the example of a round baler.

File Server (FS)

The File Server enables data exchange with external instances (e.g., USB drive or cloud connection) and ISOBUS devices. The File Server acts as a central place to store or retrieve data. It standardises the data transfer method, but the data content can, in fact, be proprietary. Usually, the File Server is located in the terminal.

One of many use cases for farmers, dealers and manufacturers is the installation of a new spreader table for a fertiliser spreader, which was not available at the time of production of the controller. Another benefit would be easy transfer of software updates to an implement controller or saving machine settings to an external drive as a backup.

Compatibility between ISOBUS systems

To ensure the highest possible degree of reliability for customers, the AEF has developed the AEF Conformance Test. It is designed to ensure that ISOBUS products actually support a specific functionality, i.e., they comply with ISO 11783 as well as with the accompanying AEF Guidelines.

A certificate is issued for each product that successfully passes the AEF Conformance Test. Certifications are required for listing products in the AEF ISOBUS Database. Five independent test laboratories can carry out the test.

Conformance Testing and Certification to guarantee compatibility

In order to manage the whole process of certification of ISOBUS components, the AEF has developed a highly automated AEF Conformance Test for its members and the AEF recognised test laboratories. The Conformance Test offers formal checking and testing of ISOBUS products by the Test Labs against the ISOBUS standard and the defined AEF functionalities.

Only When successfully passing this test, the manufacturer receives a certification, which allows to publish the AEF certified components in the AEF ISOBUS Database.

This is shown to the public by the AEF Certified Label which can be visible on equipment as seen in Fig. 3.8. The aim is a clearer description of the effectiveness of a manufacturer independent ISOBUS system and increased operational reliability for the farmer. The AEF Conformance Test tool is also available for the development departments of AEF members, to enable continuous checking for compliance with the standard during the development phase of their own ISOBUS products.

FIGURE 3.8

AEF certified label.

https://www.aef-online.org/.

AEF recognised test laboratories

AEF has currently appointed five international test laobratories that are allowed to perform the AEF certification process:

- REI — Reggio Emilia Innovazione (Reggio Emilia, Italy)
- ITC — ISOBUS Test Center (Osnabrück, Germany)
- NTTL — Nebraska Tractor Test Laboratory (Lincoln, Nebraska, USA)
- DLG (Groß-Umstadt, Germany)
- Kereval (Thorigné—Fouillard, France)

AEF ISOBUS Database

Detailed information on the certified product can be found in the AEF ISOBUS Database at www.aef-isobus-database.org. Multiple components within the same ISOBUS system (Tractor and Implement) can be selected to show the compatibility and identify the lowest common denominator (Fig. 3.9).

Many questions are now answered through the AEF ISOBUS database:

- Who is responsible if components do not work together: the manufacturer of the tractor or of the implement?
- How do I find a fully ISOBUS compatible implement for my ISOBUS tractor, in order to make use of the full functionalities of the system?
- Is my existing implement perhaps ISOBUS certified and compatible with the new ISOBUS tractor to be purchased? And if so, what functionalities can I use with the combination?

FIGURE 3.9

AEF ISOBUS Database: Compatibility check of certified systems.

https://www.aef-online.org/.

The AEF ISOBUS Database contains all relevant information of the ISOBUS-certified machines and equipment. After selecting a combination of tractor and implements with a few mouse clicks, the functionalities and the status of compatibility of the selected combination is visible. If a tractor or implement a user is searching cannot be found, it is not certified. The AEF ISOBUS Database helps retailers in advising their customers and also facilitates troubleshooting by the after sales service. This can significantly reduce downtime.

The AEF ISOBUS Database is updated continuously since it is also used for determining the conformance of machines and implements with the ISOBUS standard as well as the certification of this conformance.

AEF Plugfest

ISOBUS is a communication protocol meant to standardise the communication systems on ag equipment from different manufacturers. However, due to its complexity, the ISO 11783 standard has been interpreted by each manufacturer slightly differently. AEF Plugfests began in 2001 to help engineers test products under development, and ensure that manufactures of tractors, implements and components were interpreting the standard in a compatible way. AEF Plugfests allow engineers to connect their ISOBUS products together, checking to see if their products can communicate with each other. At the end of the event, all attendees will have new data and a deep understanding of how to improve their products (Fig. 3.10).

FIGURE 3.10

AEF Plugfest 2019 in Antibes, France.

https://www.aef-online.org/.

The Plugfest is a semiannual event held in the United States and Europe to test the protocol defined in the ISO 11783 standard. The event attracts hundreds of attendees and participants.

A great benefit of the Plugfest is that it allows the newcomers to the market to get valuable feedback on their product development regarding the ISOBUS compatibility. Plugfests are also beneficial for the established companies to test their new software generations, which are not yet released in market for farmers.

Future developments on ISOBUS

What comes next? Agriculture 4.0 and the digitalisation show lots of possibilities which challenge the Ag industry. AEF and the Ag industry have started developing new projects and work on High-Speed ISOBUS and Wireless in-field communication as well as defining interfaces between cloud solutions.

High-Speed ISOBUS (HSI) including future network architecture

Whilst there has been a perception of the need for a faster network for several years, one of the first steps was to identify and document use cases requiring a faster network.

- More precise command and control of prescription
- More precise data logging of as-applied and yield information
- Higher featured and more responsive display of information

For these and many additional use cases, High-Speed ISOBUS experts within the AEF have developed the following required attributes: communication speed, bandwidth requirements and control latency. The team then worked to scale those requirements into imagined future systems which would leverage these needs — as a bigger, better and faster combination of tractor, implements and integrated functionalities.

Task Controller is one of the functionalities that drives HSI. In looking in more detail, it elaborates on that high-level requirement:

- Command and control timing accuracy well below 10 ms
- Control of all possible set-points within that time interval
- Substantial reduction in jitter in delivering that control

Examples of future technologies using HSI will include higher performance command and control at the row level on a large planter or for the individual nozzles on a sprayer; higher performance of User Interface to see the exact state of every row in real-time, whereas today the operator may see screen updates that are somewhat lagging; remote process viewing with more advanced digital cameras replacing the analog of yesteryear, easily integrated into an HSI system; enabling higher levels of automation; improving diagnostics and faster software updates; connectivity to other in-field machines using AEF Wireless Infield Communications, and more.

Along the journey of use case and requirements development to serve these new systems, the High-Speed ISOBUS team has had deep and on-going engagement with two other AEF project teams — Camera Systems and Wireless infield communication.

Challenges: Whilst many requirements are like those in the automotive industry, others significantly exceed what has been observed so far, such as the plug and play integration of implements to tractors, as well as the overall "size" of the network from one end-point to another. Thus, even as the benefits of automotive work are translated to Ag, there are other areas where the High-Speed ISOBUS experts work appears to be leading the industry. Some years back, the experts naively thought that we could identify the technologies that will be useful, and that they will already exist — we would somewhat simply need to identify the needed protocols and the needed cable and connector components and move forward in our work. This has proven not to be the case.

Connectors for our industry have a much more challenging environment than on automotive systems. Implements are routinely connected and disconnected, and the environment is quite harsh. We are working with several connector companies towards a solution to 'cross the hitch' in a robust and reliable manner. Cabling that meets the performance needs and is applicable to our industry is also a challenge. In automotive, attention is on size and weight, and cables are affixed to the body and chassis structure to minimise flex and failure. These types of cables would not survive the Ag environment. To develop a reliable solution, High-Speed ISO-BUS members have been working with cable and harness suppliers to ensure Ag environment is properly accommodated.

Wireless In-field communication (WIC)

The diversity of the agricultural equipment market leads customers to operate heterogeneous fleets with choices for various manufacturers based on their individual preferences. As a matter of the digital evolution in the agricultural industry, connectivity and data exchange between cooperating machines are essential. This includes local information flow in terms of machine synchronisation, guidance line and coverage sharing as well as end-to-end communication of high-definition sensor data all the way through the network cloud down to analysis and management software tools. In the field domain, this data exchange fuels solutions for operator assistance and converges into fully automated workflows (Fig. 3.11).

In future, the global connectivity will be the basis for remote implementations of agronomic decisions as well as autonomous field operations. Wireless In-field Communication focus spans from technology decisions for suitable radio standards to corresponding transport layer protocols for machine-to-machine (M2M) communication and vehicle security. At the boundaries of the agricultural domain, WIC systems will be in touch with Cooperative Intelligent Transport Systems (C-ITS) in

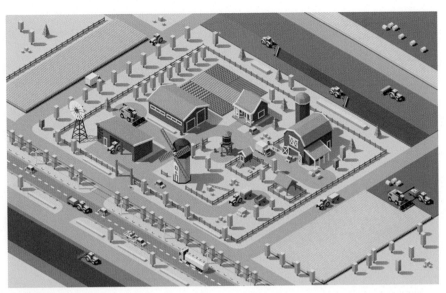

FIGURE 3.11

Wireless in-field communication use cases: Road Safety, Platooning, Process Data Exchange.

https://www.aef-online.org/.

order to provide road safety information. Based on several car manufacturers' announcements, the corresponding technology has entered the market with their upcoming models. This creates significant potential for the avoidance of severe road accidents involving agricultural equipment (Fig. 3.12).

With the work toward standardisation, this and other use cases were identified, amongst which include:

- Transfer of prescription and coverage maps from one machine to another for cooperative use
- Remote viewing of a camera on a nearby machine
- Command and control of an adjacent machine (Platooning)
- On-road and in-field safety (C-ITS)

Each of these use cases have their own bandwidth and latency requirements. Given the need for high bandwidth transfer, the inclusion of these requirements into the HSI architecture can again lead towards reduced system complexity.

With the first field tests, 802.11p as radio technology proved its suitability to be adapted to Ag use cases. Good range in combination with sufficient bandwidth and availability of different chips-sets and communication stacks combined with a usage in a wide variety of passenger and truck application make this technology today our number one choice.

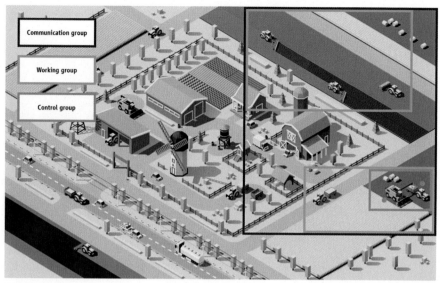

FIGURE 3.12

Different working groups based on defined use cases.

https://www.aef-online.org/.

AEF is working together with partners from universities and industry to make infrastructure free, direct and interoperable machine-to-machine communication as a major enabler of digital-driven agriculture.

Connectivity between platforms

Today almost everybody active in agriculture and food production is handling data, from the input providers like seed and chemical companies, the farmers themselves, to the customers of the farm products such as food processors, wholesalers and retailers. Not only is agricultural equipment one of the major generators of this data but these machines also require data to operate efficiently and support precision farming.

Most of the agricultural equipment manufacturers have their own software platforms which allow their customer to connect to their vehicles and view the data being generated, whilst some of the smaller manufacturers have teamed up and have joint platforms. From the customers (farmers) point of view, the availability of their data is a must, but due to the mixed fleet nature of farming operations today, it means accessing more than one software platform.

Most of the software platforms can be considered as Farm Management Information Systems (FMIS) and enable the farmers to manage their farming operations by allowing them to view and manipulate the generated data. The FMIS for the farm is similar to the ERP (Enterprise Resource Planning) for a manufacturing facility but, rather than manufacturing lines, it contains fields and vehicles. Some FMISs are more agronomic, whilst others have an operational, financial or regulatory focus and these tend to be tailored towards local needs.

The ability to manage the farm more effectively revolves around the ease of getting all of the data available for reporting, analysis and to derive insights for future operations. Whilst there are a number of solutions on the market today which enable multiple platforms to exchange data, there is still no standardised connection for the agriculture industry.

The Ag industry is working to enable multiple platform to platform connections in a standardised way, enabling farmers to get all their data in one desired location for reporting, analysis and to derive the insights that are possible and becoming more and more necessary in today's precision farming world.

Field robotics for harvesting

4

A review of field robotics approaches for harvesting

Josie Hughes[1,2], Fumiya Lida[1], Simon Andrew Birrell[1]

[1]*Department of Engineering, Cambridge University, Cambridge, United Kingdom;*
[2]*Computational Robot Design & Fabrication Lab (CREATE-Lab), Lausanne, Switzerland*

Introduction

The story of agriculture is one of increasing automation. Crops are planted, weeded and harvested with ever-decreasing direct human involvement, reducing labour costs and improving yield. However, every fruit or vegetable is different, and solutions for a single crop can vary from country to country and even company to company. Whilst some crops such as wheat or potatoes have long been harvested mechanically at scale, many others such kiwi fruit (Scarfe et al., 2009), cucumbers (Van Henten et al., 2002), citrus fruit (Harrell et al., 1990), strawberries (Hayashi et al., 2010), broccoli (Kusumam et al., 2016), grapes (Luo et al., 2016; Monta et al., 1995) and many others (Bac et al., 2014) have resisted commercial automation. Agricultural robotics presents unique challenges compared to robotics in the more common factory environments (Denny et al., 2009). Agricultural environments are unstructured, intrinsically uncertain, harsh on mechanical equipment (Reddy et al., 2016) and have high variability over weather conditions, locations and time. Autonomous agricultural systems must be flexible and adaptive (Hajjaj & Mohamed Sahari, 2016), (Edan et al., 2009) to cope. Harvesting and other crop manipulation tasks (Kemp et al., 2007; Hughes et al., 2018) are particularly challenging (Bac et al., 2014) along all these dimensions.

Field robotics for harvesting

Despite the variability in crops and robotic systems, the robotic harvesting processes can be described as a generalized process, as summarised by Fig. 4.1.

Many harvesting robotics systems use many or most of these components and processes; however, not all crops of robotic systems require all processes.

The first component is **sensing**, most often using cameras, to obtain data that provides information as to the crop and of the location of the harvesting system. Typically, this could be an RGB camera, depth camera or a lidar system. The choice of sensory system often depends on the crop or environment in which the robot operates.

Digital Agritechnology. https://doi.org/10.1016/B978-0-12-817634-4.00009-4

69

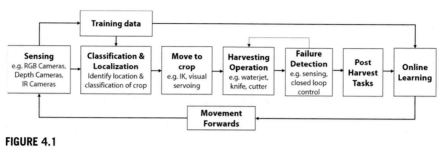

FIGURE 4.1

A generalised process for harvesting robotics.

Classification and localisation can then be performed using the sensory information to allow meaningful decisions about where and what to harvest. Localisation involves locating a specific instance of the crop, for example, a specific tomato or lettuce that requires harvesting. It is important that localisation has a precision that is compatible with the physical harvesting device. After localising the plant or fruit, it is then often useful to classify the specific instance. For example identify if the item is under-ripe, damaged, infected or ready for harvest. This helps not only to minimise waste, but also to potentially reduce the spread of infection. There are many different approaches for localising and classification, which can in some cases be a combined process. Many approaches utilise machine learning techniques, relying on significant amounts of **training data**. A key challenge for the localisation and classification process is achieving robustness to the changing conditions (e.g., lighting, weather, field or crop variability) that may occur, and also limiting the amount of training data that is required as collection and labelling can be expensive.

After identifying the location of the produce to be harvested, it is necessary to **move to the crop**. If a manipulator is being used, inverse kinematics can be used to move the end effector to the crop. In some cases, using additional techniques such as visual serving can be used to increase the accuracy. After moving to the crop, a mechanical process can then be performed, the **harvesting operation**. There are a number of different techniques including water jet, rotary cutter, scissor cutters, band saw or even suction. In some cases, in particular for crops which are challenging to pick or there may be a high failure rate, there may be a stage of **success or failure detection**. This may include using sensory data or imaging to check the item has been removed, or even to check the quality of the cut. If there is a failure, it may be possible to return and repeat the harvesting operation. In addition in some cases, there may a **post-harvest** step. This could be a sensory driven step such as grading of the crop by metric such as quality or size, or even a manipulation process, such as trimming our outer leaf removal.

After this cycle of the robotic harvesting process, there may be a step of **online learning** where sensory data, or experience, maybe be used to learn how to complete the process better. This may incorporate training data gathered earlier in the harvesting process. For example, using feedback from the harvesting operation the robot could learn improved control strategies, or learn to classify produce better. This could utilise humans-in-the-loop. Finally, the robot can **move forwards** or physically advance onto another item to be harvested and the cycle can continue.

Achieving impact: faster, cheaper and safer harvesting

There are number of ways in which robotic agriculture can be applied to achieve significant impact. Impact must be shown to achieve positive adaption from the agriculture communities. It can be shown in a number of different ways, through faster, cheaper or safer harvesting.

Faster harvesting

Agricultural robotics could be used to increase the throughput that can be achieved in comparison to manual processes. This could be used to allow demand to be met quicker and could be used to address the increasing problem of limited labour. Robotics may be able to sustain or achieve faster harvesting with reducing labour.

Cheaper harvesting

The introduction of agricultural robotics could assist with reducing costs associated with harvesting. This could be through a number of mechanisms. Firstly, reducing labour costs. Although robotic systems may have a greater upfront costs, over many seasons they may offer reduction in the number of workers required. Secondly, reduced waste or damage; through increasingly intelligent harvesting approaches, the amount of produce which are wasted or damaged unnecessarily could be reduced. Thirdly, the improved access to data or single plant tending could lead to better crop yields and performances, increasing the quality and hence profitability of the crops.

Safer harvesting

Agricultural robotics could reduce the need for workers to complete back-breaking, potentially dangerous harvesting work, reducing the dangers faced, and their exposure to harsh conditions. In addition, reducing the number of works around the harvesting process could increase the safety of field environments.

Within agriculture there are different scenarios or crops where robotics could be deployed most effectively. One of the key challenges for agritech is identifying of the key crops where impact can be made in one of these three ways. This combines both crops and environments where automation can be particularly helpful, and also in scenarios where the technology that can be applied is of sufficient maturity.

A final and increasingly critically potential impact of agricultural robotics is the reduced environmental impact, and contributions to achieving greater sustainability.

Increased sustainability

With sustainability becoming an increasingly pressing issue, robotic harvesting has the potential to help reduce environmental damage or harm caused in the harvesting process. Robotic harvesting could reduce the damage to soil which occurs from heavy traffic of heavy farms vehicles. In addition, more efficient harvesting could reduce waste of damage making the process more efficient in terms of the use of pesticides or herbicides.

The reality of harvesting

Field robotics and agricultural robotics are challenging not only because of the realities and harshness of the physical environment, but also the reality of the economic models and user acceptance.

Challenges

Firstly, there are a number of technical challenges that agricultural robotics must consider.

- **Harsh Environments.** Agricultural fields can exhibit a range of harsh conditions from low temperatures and heavy rain, to baking temperatures and high humidity. Physically, robots must be able to operate in these conditions without any down time.
- **Variability and Uncertainty.** Ever field, plant and crop is different, and every day brings different weather conditions. The robotic system must be able to deal with this variable and uncertainty.
- **Speed and Quality.** Harvesting must be performed rapidly and too high quality standards dictated by the supermarkets. It requires precision and accuracy in all aspects of the harvesting process.
- **Unstructured Environment.** The field is very loosely structured, with few known or structures. This makes it challenging to make references points or use or develop accuracy calibration routines.
- **Reliability and Maintenance.** Robots must work reliability and with minimal maintenance for long periods of time.
- **Human-Centred Environment**. Agricultural conditions are very human centric and required human workers. Robots must be safe in such conditions, and pose no danger to workers who may be close by.

In addition to the technical challenges, there a number of wider challenges to the agricultural robotics operation:

- **Economically Challenging.** Agricultural produce have very small margins. Therefore, any solution must be economically viable and cost effective.
- **Slow-moving Industry.** Agriculture has often been seen as to be a slow-moving industry which is slow to adopt new technologies and techniques.
- **Robots taking human jobs.** One potential impact of increased automation is the loss of jobs. Although such jobs may be hard and back-breaking, they provide many with work. The media-driven perception of robots taking over human worker jobs is hard to address.

- **Fixed Environments and Infrastructure.** Much of the environment and infra-structure is fixed, for example, field layouts, of standard sizes of farm vehicles. There is limited ability to change of vary the environment in the near future.

Current state of the art

There has been much research and development of harvesting robotics in the last 20+ years. The work has focused on a variety of different crops and uses different methods and technologies. Table 4.1 summarises the different field robotics approaches, for different crops including a summary of the harvesting approaches and also the computer vision learning approaches. In particular, we focus on developments in the last 10—15 years.

This chapter demonstrates the wide variety of approaches required for the different crops. In the following sections, we review on-going research and current state-of-the-art approaches for the different components of harvesting.

Table 4.1 A generalised process for harvesting robotics.

Robot system	Year	Harvesting mechanism	Sensing	Learning	Level of development
Cherry robot (Tanigaki et al., 2008)	2008	Suction	Scanning 3-D vision sensor	Segmentation for detection	Limited field tests
Iceberg lettuce (Simon Birrell et al., 2019)	2020	Belt, pneumatic driven knife	RGB webcams	CNN localisation and classification	Field tests
Cucumber robot (Van Henten et al., 2002)	2002	Thermal cutting	3D imaging system	Detection and 3D imaging	Greenhouse tests
Apple robot (Zhao et al., 2011)	2011	Electric cutter	Vision sensors, position and collision sensors	SVM classification	Field tests
Strawberry robot (Hayashi et al., 2010)	2010	Suction and pneumatic cutting	CCD cameras	Detection and classification	Field tests
Tomato robot (Monta et al., 1998)	1998	Suction cups	—	—	End effector only

Continued

Table 4.1 A generalised process for harvesting robotics.—*cont'd*

Robot system	Year	Harvesting mechanism	Sensing	Learning	Level of development
Radiccio robot (Foglia & Reina, 2006)	2006	Pneumatic mechanics and muscles	CCD camera	Thresholding for localisation and detection	Some field tests
Fruit robot (R Ceres et al., 1998)	1998	Cutting gripper	Laser range-finder	Localisation	Lab tests
Apple robot (Bulanon et al., 2004)	2004	–	Binocular stereo vision	Unitary linear regression	Component tests
Sweet pepper (Hemming et al., 2014)	2014	Fin ray gripper and cutter	Colour 3D vision	Localisation and obstacle avoidance	Autonomous greenhouse tests
Tomato (Kondo et al., 2010)	2010	Custom cutter	Tactile strain sensors	–	Greenhouse tests
Kiwi (Scarfe et al., 2009)	2009	4-arms with custom gripper	8+ webcam	Stereopsis	Field tests
Eggplant (Hayashi et al., 2002)	2002	Custom gripper-cutter	Cameras	Segmentation and division	Some tests
Orange harvester (Armada et al., 2005)	2005	2005	Custom grippers	Vision and learning	Field tests and Simulation
Cauliflower (FB Klein et al., 2019)	2019	Variable stiffness actuator	Intel realsense	Classification	Lab tests

Mechanisms and manipulation

As demonstrated in Table 4.1, there are a number of different approaches to the physical harvesting mechanism and mechatronics of the harvesting process.

Blades

Blade-based cutting mechanism is often used. A variety of different actuation methods are used to achieve the dynamic profile necessary for cutting, with the choice of actuator depending on the harvesting action required. Often significant force and quick rapid movement along a single plane can be required to achieve neat 'supermarket' ready cuts. As such, typical methods for actuation include pneumatic actuators, belt drive, band saw or scissor mechanisms.

Waterknife

A relatively new concept, waterknives, uses high pressure water to 'cut' through matter, similar to a waterjet cutter. Unlike typical blades, the water jet method does not show wear or contamination. However, it can be hard to control or position and requires significant amounts of high pressure water. Although it is not appropriate for all crops, it offers a promising cutting mechanism for a number of crops.

Suction

Suction is used in a number of modalities. Firstly, suction cups, often soft, are used for gripping and manipulation of single fruit or produce. Often this is combined with a secondary manipulation mechanism. This is often a good solution for soft easily damaged fruit. Suction is also used to fully remove more robust fruit from the plant. In particular, this has been seen for apples which are literally sucked of the trees.

In-hand manipulation

With advances in robotic manipulation and soft robotic technologies, the capabilities of manipulators are increasing. This allows for more complex manipulation tasks, which may be more anthropomorphic in inspiration. For examples, grippers which can surround fruit or delicately remove them from the plant. This also allows for other manipulation tasks which must be undertaken as part of the harvesting process — for example, some post-process or tidying processes such as leaf removal or stem removal. Although this approach is more complex, it allows for more bespoke and complex interactions.

Vision and learning

Harvest localisation and classification

The application of vision and learning techniques in agricultural settings dates back to the 2000s, before the widespread use of neural networks (NNs), where a wide variety of hand-crafted features and ad-hoc vision solutions were usually used. These technologies were often developed with the objective of automatising the harvesting industry; thus, solutions were often designed to function within robotics harvesting systems, where the localisation and classification of single produce within crop fields was necessary.

In Nieuwenhuizen et al. (2010), the detection of volunteer potato plants was performed using adaptive Bayesian classification of Canny Edge Detectors amongst other features. Broad-leaved dock detection (a weeding task) was performed using a texture-based approach, where image tiles were subjected to a Fourier analysis (Evert van et al., 2011) (Weeding is a similar task to harvesting, just with less concern for the fate of the extracted plant). An alternative approach to weed detection used wavelet features of Near Infrared (NIR) imagery (Scarfe et al., 2009), subsequently passed to a Principal Component Analysis (PCA) component and a k-means classifier (Kiani et al., 2010). Grapes have also been detected with Canny Edge filters, using Decision Trees as the classification mechanism (Berenstein et al.,

2010). Foliage detection on the same project required a separate algorithm. Grapes were classified on another project using the AdaBoost framework, which combined the results of four weak classifiers into one strong one (Luo et al., 2016). Radicchios have been detected by thresholding Hue Saturation Luminance images and applying particle filters (Foglia & Reina, 2006). Cucumbers were detected using NIR photography at two positions 5 cm apart, to give stereoscopic depth information (Van Henten et al., 2006) and classified for maturity by estimating their weight from the perceived volume (Van Henten et al., 2002).

The use of ad-hoc vision techniques to solve these problems presents several advantages, pre-eminently amongst which transparency of the detection procedures and the possibility to appropriately assess failure. This, like several additional reasons, made it so that tailored vision solutions in agricultural settings are still in use. A more recent experiment, for example, detected broccoli heads using an RGB-D sensor, had the disadvantage that the robot had to move a tent across the field to prevent interference from outdoor light. Point clouds were clustered from the depth information, outliers were removed and Viewpoint Feature Histograms constructed. A Support Vector Machine performed the actual classification of the broccoli heads (Kusumam et al., 2016). The use of vision to provide control through methods including visual serving has also been shown to increase positional accuracy when harvesting citrus fruit (Mehta & Burks, 2014; Mehta et al., 2016). Ad-hoc vision techniques for pose estimation were developed (Hughes et al., 2018), where lettuce pose estimation was performed through a series of visual tests locating the lettuce within the visual field of the camera, and its orientation based on stem localisation.

With the success achieved by NN architectures in the field of vision, it is not surprising to have observed an increase in NN-based solutions for detection and localisation of crops in the past 2 decades (Kamilaris & Prenafeta-Boldú, 2018). In a study by Han Lee et al. (2015), for example, the deep-plant system based on a Convolutional Neural Network (CNN) architecture was shown to be able to accurately differentiate between 40 different types of plants by venations patterns. Later, in a study by Han Lee et al. (2017) a new Deep architecture (HGO-CNN) was shown to outperform its predecessors. In a study by Habaragamuwa et al. (2018), a Deep CNN architecture was proposed for the detection of strawberry fruits, whilst in a study by Ge et al. (2019), a similar deep architecture was augmented with 3D localisation features for robotics strawberry picking, including safe region pickable strawberry identification. In a study by Simon Birrell et al. (2019), an end-to-end NN architecture based on the YOLO detection system (Joseph & Ali, 2018) was shown to be able to locate and classify iceberg lettuce in the field. Some of the disadvantages of these solutions lie with the need of large amount of labelled data, which is usually expensive to both gather and label in agricultural settings. Moreover, as Deep Neural Network-based architectures lack transparency and readability, it is often hard to appropriately assess and understand failure. In the context of a full system, misdetection at an early harvesting stage may have severe repercussions at a later stage. Understanding and appropriately addressing failure thus becomes a necessity.

Finally, some of the best performing solutions see the merging of both worlds, where hand-crafted features and ad-hoc vision techniques for image pre-processing improve the performance of NN-based architectures. In a study by McCool et al. (2016), for example, sweet peppers were localised and detected by first performing pixel-level image segmentation and second using a Laplacian of Gaussian multi-scale blob detector to classify the segmented image. The segmentation was performed by a combination of Histogram of Oriented Gradients (Dalal & Triggs, 2005), Local Binary Patterns (Ojala et al., 2002) and Sparse Auto-Encoder architectures (Ng & et al., 2011; Hung et al., 2013). In a study by Inkyu et al. (2016), image pre-processing by highlighting regions of interests was shown to improve classification (Kamilaris & Prenafeta-Boldú, 2018). In a study by Chen et al. (2017), a system for counting apples and oranges with a deep NN architecture was improved by the early detection of bounding boxes. Other work, instead, aimed at augmenting CNN performance via statistical features, histograms or PCA transformations (Rebetez et al., 2016; Wang & Cheng, 2015).

Crop monitoring

Precision agriculture comprises a set of technologies that combines sensors, information systems, enhanced machinery and informed management to optimise production by accounting for variability and uncertainties within agricultural systems (Gebbers & Adamchuk, 2010).

In the context of the precision agriculture of the future, whole field crop monitoring systems have become a steady direction in the development of new technologies. This direction has been facilitated by the rapid advancement of Unmanned Aerial Vehicles (UAVs) technologies, with cheaper devices with better sensors, longer flying times and better AI technologies. It is in this context that Deep Learning framework for aerial vehicles in the context of agriculture has been developed. In a study by Bah et al. (2018) and Li et al. (2016), for example, a Deep CNN architecture was developed to identify weeds via UAV imagery. Others systems, instead, focused on disease identification, like the Fusarium wilt of radish (Ha et al., 2017) or vine diseases (Mohamed et al., 2018). Finally, usage of UAV and deep learning techniques for higher crop throughput or to analyse yield and fertilisation levels has been explored (Ampatzidis & Partel, 2019; Escalante et al., 2019).

On the opposite side of UAV monitoring systems is specialised crop monitoring pest surveillance and detection, which usually requires on-the-ground systems with high-resolution sensors on single monitored crops. In Li et al. (2019) and Yue et al. (2018), for example, deep learning pipelines were proposed to localise and count agricultural crop pests.

Post-harvest and quality control

The quality of crop is dependent on several pre-harvest factors, amongst which are weather conditions, growing land, irrigation patters, chemical treatments and more besides (Sams, 1999). In this context, during and immediately after harvesting, quality control partakes a fundamental role. When harvesting, in fact, it is often the case that growers necessitate careful inspection of produce. The specific inspection of agricultural items can inform the grower of the time of harvest, with ripe produce needing earlier harvesting, the need of additional substances like chemicals for weed or pests control or withering produce necessitating additional watering.

Quality control usually includes inspections for size, colouring, shape, maturity disease and other factors. One of the most important characteristics influencing quality, and the marketability of produce, is ripeness (Kader, 1997). Besides appropriate harvesting time, determining the ripeness of horticultural produce is useful for classification, transportation, handling and the security of its quality. Spectral techniques have commonly used to assess the quality of post-harvested produce (Hussain et al., 2017; Jha et al., 2010; Slaughter, 2009, pp. 1–18). More specifically, Raman imaging, Fluorescence imaging, Laser backscattering imaging, Hyperspectral imaging and Nuclear magnetic techniques have been shown to be able to classify produce lacking chromaticity differences into various stages of ripeness. The equipment required for said methods, however, is usually bulky and the information processing is often computationally intensive, making it hard to create solutions which can be exported in the field, or do not require the transportation of produce to appropriately equipped areas.

Deep learning systems have been shown to work well in quality control settings, with camera-based systems demonstrating the possibility to assess the ripeness of horticultural produce by visual cues (Mustafa et al., 2008; Shah Rizam et al., 2009; El Hariri et al., 2014; Eyarkai Nambi et al., 2015). These systems, mainly detect differences in chromacity at various ripeness stages for their maturation assessment. Other visual inspections are also possible. In a study by Arakeri et al. (2016), for example, a vision-based system for quality control of tomatoes is proposed. In the system, colour statistical and texture features are first extracted, before a feed-forward NN is used to perform classification based on defects and ripeness. Crop disease recognition is a recurrent topic for many crops, including Deep CNN solutions for disease recognition of citrus (Dong et al., 2019), plant leaf disease (Hanson et al., 2017) and apples (Moallem et al., 2017). In the latter, delicious apples are graded based on surface features with a combination of NN, Support Vector Machine and Nearest Neighbour classifiers for the detection of defected produce.

Learning, however, is not solely based on visual sensing. When assessing ripeness, for example, consumers use a combination of tactile sensing and visual cues. The physical probing of produce is one of the oldest modalities for ripeness assessment, and brought the advent of penetrometer measurements (Blanpied et al., 1978), not the least because of the instrument's ease of use and transportability. Technological advances in tactile sensing and perception (Nicolas et al., 2018; Culha et al., 2014) have changed the landscape for tactile-based inference procedures (Pfeifer et al., 2014; Scimeca et al., 2018). Recent work by Scimeca et al. (2019), for example, proposes a non-destructive automated solution to ripeness estimation of mango fruits, with a robotic end-effector equipped with a capacitive tactile sensor and performing a light touch on the produce.

Field tests and measuring success

To allow comparison between different setups, and also to provide methods of quantifying, the performance there is a need to have common performance metrics across field robotic systems. These must also map to figures which are commonly used within agriculture, and are directly correspond to economic or performance factors. In particular, these were summarised and introduced by Bac et al. (2014); a number of previous works adhere to these metrics to allow for the comparison of **radicchio** (Simon Birrell et al., 2019). We summarise a number of these metrics in Table 4.2, along with example values for the lettuce-picking Vegebot (Simon Birrell et al., 2019) and a melon harvester (Edan & Miles, 1993). Their significance is now discussed.

Target Localisation Success and **False Positive Target Detection** measure the overall performance of the classification and localisation system. Both metrics have as a denominator the number of real qualified fruit or vegetables. 'Qualified' is binary, so further potential classifications such as 'Immature' or 'Diseased' are not profiled. Nevertheless, these metrics give a good first approximation of performance and allow comparison between widely varying systems. Notice that the Vegebot has

Table 4.2 Definition of metrics and example values for two case studies: lettuce and melons.

Metric	Calculation	Lettuces example (Simon Birrell et al., 2019)	Melons example (Edan & Miles, 1993)
Target localisation success	$\dfrac{\text{Number of detected qualified targets}}{\text{Number of real qualified targets}}$	88%	94%
False positive target detection	$\dfrac{\text{Number of false positive detections}}{\text{Number of real qualified targets}}$	1%	20%
Detachment success	$\dfrac{\text{Number of successfully harvested targets}}{\text{Number of localised targets}}$	97%	92%
Detachment attempt ratio	$\dfrac{\text{Number of detacment ttem ts}}{\text{Number of successfully detaced targets}}$	1.0	1.3
Damage rate	$\dfrac{\text{Number of damaged targets}}{\text{Number of localised targets}}$	38%	7%
Cycle time	Average end-to-end time for fully locating and harvesting a target, including failed attempts.	32s	15s
Number targets evaluated	Total number of targets evaluated in calculating the other metrics.	69	400
Harvest success	Target localisation success × detachment success	88%	86%

a poorer localisation success than the melon harvester, yet admits fewer false positives; this is often a trade-off that can be tuned in the classification and localisation system.

Detachment Success measures the ability of the system to detach fruit or vegetables from the set of targets it has previously detected, isolating the performance of the harvesting unit from the classification and localisation system. **Detachment Attempt Ratio** measures the average number of attempts made on each successfully detached target, qualifying the overall detachment success.

Not all successfully detached targets are in market-ready conditions, so the **Damage Rate**, which is the proportion of detached targets that are not saleable, contributes to an economic evaluation of the device. The Vegebot has a poor damage rate of 38% compared to the melon harvester's 7%, largely as a function of the stringent aesthetic requirements of supermarket chains for iceberg lettuce.

The **Cycle Time** allows the calculation of the fruit or vegetable that can be harvested per device per day and is of key economic importance. It measures the performance of the overall device, from target detection to harvesting to motion through the field. It also allows a direct comparison with the speed human harvesters. The Vegebot's cycle time is 32s, compared to the human cycle time of 7s.

The **Number Targets Evaluated** gives an indication of how statistically significant all the other cited metrics are. **Harvest Success** is another overall performance metric, derived by multiplying the Target Localisation Success by the Detachment Success.

Each project will have more specific metrics, but these allow a comparison of the state of the art for different crops.

Translation from research to commercialisation

A 2013 survey paper noted that after 30 years of research into harvesting robotics, none of the 50 robots surveyed had reached the stage of commercial exploitation. That situation has changed in the past few years, with the emergence of start-ups focused on building commercial products.

The key driver of demand for robotics harvesters is the uncertain availability of human labour (Colin, 2019). Immigrant labour is harder to obtain, both financially and politically, and the population of advanced economy show little enthusiasm to work on the land. In addition, just in time supply of produce to supermarket chains requires that farmers must be able to marshall harvesting resources at very short notice.

Commercial **strawberry**-picking robots include those manufactured by Dogtooth in the United Kingdom (Youngman, 2017), Octinion (Belgium) and Harvest CROO (USA) (Fig. 4.2). The Dogtooth and Octinion devices have single robotic arms on mobile platforms, whilst the Harvest CROO uses a large mobile platform that picks many berries in parallel (Staff, 2019). **Tomato** harvesters are being commercialised by Metomotion (Israel) (Siegner, 2019) and Four Growers (USA). These devices have the advantage that tomatoes are vernally grown in the

FIGURE 4.2

Examples of commercially developed harvesting robotics. Clockwise: dogtooth robotics strawberry harvester, octinion strawberry harvesting, metomotion harvester and fieldwork robotics raspberry harvester [ADD REFS].

less challenging environment of a greenhouse. Other crops have lone entrants tackling the problem. Fieldwork Robotics (UK), a spin-out of Plymouth University, is developing a **raspberry**-picking robot (Williams, 2019). Abundant Robotics (USA) (Gossett, 2019) use suction tubes on a mobile platform to harvest apples (Benson, 2019). AvL Motion (Netherlands) produce a large robot that can harvest 16 **asparagus** per second.

Beyond the engineering problems in producing a robust robotic harvester, there are economic and **business** challenges. Farmers typically operate on low margins, and the capital expenditure required to purchase one of the new breed of robotic harvesters makes the return on investment uncertain. Some start-ups attempt to overcome this by offering Robots-as-Service, where the producer rents the device, paying only for the time employed or the crop harvested.

Each crop requires a tailored mechanical and software solution, meaning that the above start-ups are largely focused on a **single niche crop**. One start-up that is attempting to tackle multiple crops with a single device is Tevel (Israel) that uses tethered drones to pick multiple fruits, including apples, oranges, peaches, pears and avocados. Another company, Saga Robotics (Norway), manufactures a multi-purpose mobile platform called Thorvald, with the intention that other companies can produce crop-specific tools to mount on it.

Most of the start-ups listed above use a **form factor** of a single picking end effector mounted on a mobile platform. In contrast, the products by AvL and Harvest

CROO are massively parallel and closer to tradition mass agriculture equipment. Other non-harvesting start-ups change the nature of the growing environment to ease automation. Bowery Farms (USA) has moved crop production indoors, where the environment can be precisely controlled. On a smaller scale, Farmbot (USA) places vegetables within a raised vegetable bed where a Cartesian robot has easy access. Other companies are prototyping hydroponics. Whilst these start-ups are not yet focused on harvesting, precisely controlling the environment is likely to lead to simpler picking devices.

Case studies
Vegebot: iceberg lettuce harvesting

Iceberg lettuce is an example of a crop that is still harvested by hand using a hand-held knife, and presents two main challenges to automation. First, visually identifying the vegetable's location and suitability for harvesting in what appears to be a sea of green leaves is hard even for humans. Any solution must be robust to the variation in individual lettuces, with their appearance varying greatly over weather conditions, maturity and surrounding vegetation. Second, in a terrain with an uneven ground, the lettuce stem must be cut cleanly at a specified height to meet commercial standards, whilst the lettuce head can easily be damaged by unpractised handling. A lettuce harvesting solution should therefore incorporate a high-precision, high force cutting mechanism whilst being capable of handling the vegetable delicately. There is a growing need for automated, robotic iceberg lettuce harvesting due to increasing uncertainty in the reliability of labour and to allow for more flexible, 'on-demand' harvesting of lettuce (Bechar & Vigneault, 2016).

System architecture

The system developed for autonomous iceberg lettuce harvesting (Vegebot) is shown in Fig. 4.3. Vegebot comprises a laptop computer running control software, a standard 6 degrees of freedom (DOF) UR10 robot arm, two cameras and a custom end effector, all housed on a mobile platform for field testing. A block diagram showing the integration of the system is shown in Fig. 4.4.

Vegebot contains two cameras: an *overhead camera* positioned approximately 2 m above the ground and another *end effector camera* mounted inside the end effector. Both are ordinary, low-cost USB webcams and stream video to the *control laptop*. Together, these allow Vegebot to detect (localise and classify) lettuces, and to move the end effector into position. There are additional sensors built into the *robot arm*: the standard *joint encoders* and a *force feedback sensor* that records the force and torque being applied to the end effector.

The UR10 arm provides a wide range of movements and provides force and torque information allowing force feedback to be implemented. A commercial implementation would likely have simpler arms each with an end effector, all operating in parallel (for an example of such a system, see scarfe2009development). The control

FIGURE 4.3

The Vegebot harvesting system, shown undergoing field experiments (Simon et al., 2019).

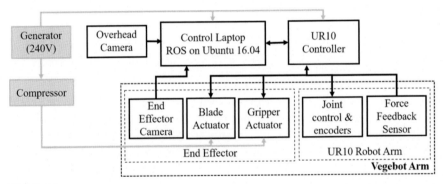

FIGURE 4.4

Block diagram of the robotic lettuce harvester system developed (Simon et al., 2019).

laptop controls the *end effector* using two digital I/O lines routed through the UR10 arm. These switch the two pneumatic actuators on and off, the *blade actuator* causing the blade to slice through the lettuce stalk and retract, whilst the *gripper actuator* causes the soft gripper to grasp and release the target lettuce.

The mobile platform supports the above hardware items and is moved manually around the field. The system is powered by a generator, which provides sufficient power to meet the peak demands of the system. An air compressor is used to enable actuation of the pneumatic systems. The generator and compressor can sit on the Vegebot to allow the system to be completely mobile.

Vision and learning

Instead, the solution chosen was to divide the pipeline into two networks (see Fig. 4.5), each trained by one of the existing datasets. The first network, a YOLOv3 object detector, would be used simply to discover the presence and location of

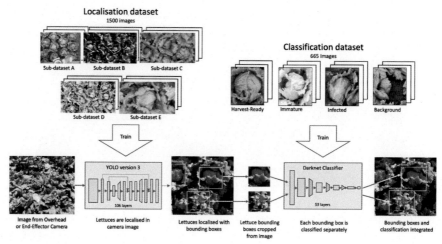

FIGURE 4.5

The vision system pipeline showing the two stages of convolutional neural network. First, the lettuces are localised using one network. A second network using both the lettuces localised from the first network and pre-segmented lettuce images from a classification dataset is used (Simon et al., 2019).

lettuces (the number of classes being reduced to a single 'lettuce' class) and output their bounding boxes. Narrow bounding boxes, likely caused by lettuces at the edge of the viewport and out of reach of the arm, are rejected as candidates. Each of the remaining bounded boxes is then cropped (adding a small margin round the outside of the bounding box to provide more visual information to the next stage) and then a second Darknet Object Classification Network was applied to each. Finally, bounding boxes predicted by the first stage and the class labels predicted by the second stage are merged. Although requiring a two-stage network, this approach offers greater performance of both localisation and classification. The architecture has been chosen to achieve the best performance with the datasets available and given the information content of those datasets.

End effector design

The end effector used only two actuators, one for grasping and one for cutting to enable simple control. A timing belt system was used to transfer the linear motion from a single actuator to both sides of the blade to allow smooth movement. This allows the actuator to be mounted above the height of the lettuce, such that when cutting it does not interfere. The belt drive system allows for the height of the cutting mechanism to be easily altered by changing the height of the cutting mechanism.

Field tests

10 experimental sessions were carried out in the harvesting seasons in 2016—18 in lettuce fields in Cambridgeshire, UK, in varying weather conditions and across many

(over 10) different fields. In these field trips, the system was developed and tested.[1] Field experiments were undertaken to test the performance of the localisation and classification system in isolation from the harvester. The entire system was also integrated to test the full functioning of the system in conjunction with its physical harvesting abilities. In this section, the localisation and classification is presented for both individual and system level tests, after which the harvesting system results are presented.

Experimental results

The results of the field experiments are shown in Table 4.3. Considering all the harvesting attempts, the detachment success was found to be 52% (31 out of 60 lettuces correctly identified, excluding false positives). However, in 28 cases, the harvesting failure was due to practical restrictions (weight of the arm, practical workspace of the robot arm and the range of the overhead camera viewport), such that it was physically not possible to pick some lettuce. If the limitations of the arm are ignored, and the denominator reflects only those lettuces within the practical workspace, then the Detachment Success rises to 97% (31 out of 32). In other words, with one exception, if the arm could reach the lettuce, the end effector could pick it. Although this is a considerable exception, it could be simply achieved by using a robot arm with increased torque output.

Table 4.3 Overall system performance in the harvesting tests. Total lettuces attempted considers only lettuces within restrictions imposed by arm strength.

Metric	Result	Definition
Total ground truth lettuces	69	
Total lettuces detected	61 (1 false positive)	
Total lettuces attempted	32	
Total lettuces detached	31	
Detachment success	97%	$\dfrac{\text{Number of successfully icked qualified}}{\text{Number of detected qualified}}$
Harvest success	88%	(Lettuce localisation success) x (detachment success)
Cycle time	31.7s, $CT^2 = 32.6$	Complete cycle time from lettuce to next
Damage rate	38%	$\dfrac{\text{Number of lettuce arvested in unsaleable condition}}{\text{Total number arvested}}$
Leaves to be removed	0.75, $CT^2 = 1.42$	Average leaves to be removed to achieve scaleability
Total lettuces attempted	69	

---

[1] These were in collaboration with a major agricultural company, G's Growers.

The future

To achieve this goal of faster, cheaper and safer harvesting, there are a number of driving technologies and also new approaches which are enabled by these new technologies.

Future technologies

Data-driven approaches and machine learning

One of the main drawbacks of learning systems lies with their biggest strength, i.e., learning by labelled examples. Like shown in previous sections, state-of-the-art supervised learning has improved automated crop detection, localisation and inspection at an unprecedented pace. The solutions, however, are often based on Machine Learning fits to large datasets, which may or may not capture the complexity of the task. When detecting iceberg lettuce (Hughes et al., 2018), for example, the variability of the produce, even within the same species, makes it hard to build solutions which are reliable against natural diversity. Moreover, if systems are expected to be deployed in real-world settings, then they must capture the variability of produce and environmental conditions within seasons, years, locations and other factors. This variability can in part be captured by the collection of large, diversity inclusive, datasets, but as the conditions of crops change in unprecedented ways over the years, even this solution would be void. If a system were to be used over decades, for example, changes would be large enough that entirely different solutions would have to be built on a case-by-case basis.

The learning necessary to solve solutions in field robotics agricultural environments is one which is adaptive, with non-stationary solutions. If a deep learning framework, for example, is made to detect fruit produce reliably during summer, the same network should ever so slightly re-adapt its solutions with the changing season, accounting for the different lighting conditions which it entails, as well as changes in colouring, size, texture and location of the fruit produce. To achieve this, learning on streaming data is thought as one of the candidate solutions, where the Machine Learning modules are retraining on new samples after deployment. These, however, may not give performance guarantees, as a change in the fit solutions may move away from well generalised pre-trained performances. In this context, it is fundamental that a system works with human operators, capable of supervising, testing and assessing the system as it improves its ability to perform detection, recognition or crop inspection. The tightly joint collaboration of human and machines can be a strong future direction for these technologies.

Finally, as the sensing learning landscape is changing, future systems should combine different sensing modalities within integrated hardware and learning technologies, which can be expected to perform delicate produce inspection, as well as manipulation. In the context of ripeness assessment, for example, tactile sensing was shown to work reliably on mango fruits (Scimeca et al., 2019). Other systems might include integrated sensing and actuation for manipulators using soft materials and/or 3D-printed technologies.

Soft robotics and tactile sensing

In recent years, there has been significant development of soft robotics technologies, and in particular, the development of soft robotic manipulators. This research trend investigates integrating softer, more compliant materials into robotic systems to enable more physically intelligent and adaptive environmental interactions. This could enable the handling of delicate or soft produce which are easily damaged, and also utilise the compliance to minimise the control complexity for highly complex manipulation tasks. This is promising for a number of delicate tasks, and also for complex manipulation tasks throughout the agricultural pipeline. Alongside this, there has also been an accompanying development of tactile sensors, which provide feedback as to the force or pressure experienced on a surface. This is another technology which could be integrated to help achieve more intelligent interactions for complex or delicate manipulation tasks.

These technologies could be integrated to allow the harvesting of delicate or challenging crops, which could not be harvested using traditional robotics approaches. There have already been a number of notable demonstrations of the use of soft robotic technologies in agricultural systems; however, there are challenges with enabling soft robotic systems which have the lifetimes require.

In addition, soft robotics offers a number of additional possible advantages. Soft manipulations could be inherently safer around human workers. There is also the capability for the compliance to offer robustness to the environment, offering long life. Solutions using soft robotics can also reduce the precision of vision or sensing systems, as the physical compliance can withstand some lack of precision. Finally, soft robotics could in some cases offer more cost effective solutions, with the compliance offering reduced complexity in the control or actuation.

Genetic engineering and bio-hybrid systems

Genetic engineering is also a developing technology that can aid robotics by developing crops that are easier to harvest. For example, by growing lettuce which grows more vertically, this could make them easier to harvest. Another related technology is that of **bio-hybrid** systems. Biological and artificial robotic systems can be interfaced together. At the plant level, this could be used to create 'super-plants'. This could also be to create robots such with worm-like functionalities to aid and improve the environment.

New approaches to harvesting

Single plant tending and harvesting

With the development of autonomous harvesting, this could enable a move away from large-scale harvesters which in a single pass harvests every plant. A new model where there are multiple smaller robotic harvesters could emerge. With the advantages of crop classification and localisation, it could also be possible to tend to, or harvest individual plants. This could allow harvesting at the optimum point of ripeness for every single plant. This has the potential advantages of minimising waste and resource usage as every item of crop is picked at the optimum point and thus there is far reduced waste. In addition, the use of smaller, lighter vehicles reduces the damage to the soil, with the potential to increase the quality of the cop produced.

24 h harvesting

Currently, harvesting is mostly limited to during daylight hours to accommodate human workers. The introduction of increasing automation could allow harvesting to be undertaken at any time of day. This has a number of potential advantages:

- Increasing the volume that can be harvested by harvesting for longer over a 24 h period
- Allowing crops to be picked at night where there may be lower temperatures, potentially increasing the shelf life of produce (Slack & Hand, 1983; Fairbank et al., 1987)
- Greater flexibility to respond to demand as harvesting can occur at any time of the day

Human-in-the loop

Although fully autonomous harvesting has the potential to offer the most advantages to agricultural processes, one stepped approach to achieve increasing automation with reduced technological challenges is implementing harvesting systems with a human-in-the loop (Hess, 2018). In this way, part of the harvesting process can be automated, but a human can still be used to provide some decision-making, validation or oversight of the automation. Some example implementations could be a human supervising multiple robotic systems, and supervising or authorising harvesting decisions.

The hybrid methodology couples the strengths of robotic harvesting with the adaptability and decision-making of humans, and provides a method of gradually increasing the automation of harvesting. This can significantly reduce the number of humans required, whilst also using a human to assist and assure reliability and safety of the automated robotic harvesting system. This is a realistic and most likely first step for the adoption of robotics for harvesting applications.

Transparency

With robot performing harvesting, we will be able to obtain more data about crops. This will make the whole operations more transparent for growers, supply chain operators and eventually consumers. At the grower level, we can better understand crops and plants, and can better meet and respond to supply chain demands. Consumers could have the opportunity to trace their produce from seed to plate, providing food security information and more information about the produce they are eating.

Structuring the environment

In the future, farm environments could be designed with robots in mind to make navigation, locomotion or manipulation easier. Already explored approaches include setting up rows or tracks to allow robots to be guided along raspberry or strawberry planters. However, the environment could be structured drastically to aid robots. For example, vertical growing systems to remove the need for locomotion, re-structuring fields with easier to access rows or growing more plants in racks. As agricultural harvesting systems become more of a reality, we must begin to also reconsider the design of the environment.

Acknowledgements

We would like to thank G's Growers and AHDB for their support in making this review possible.

References

Ampatzidis, Y., & Partel, V. (2019). UAV-based high throughput phenotyping in citrus utilizing multispectral imaging and artificial intelligence. *Remote Sensing, 11*(4), 410.

Arakeri, M. P., et al. (2016). Computer vision based fruit grading system for quality evaluation of tomato in agriculture industry. *Procedia Computer Science, 79*, 426–433.

Armada, M. A., Muscato, G., Prestifilippo, M., et al. (2005). A prototype of an orange picking robot: Past history, the new robot and experimental results. In *Industrial robot: An international journal*.

Bac, C. W., J van Henten, E., Hemming, J., et al. (2014). Harvesting robots for high-value crops: State-of-the-art review and challenges ahead. *Journal of Field Robotics, 31*(6), 888–911.

Bah, M. D., Dericquebourg, E., Hafiane, A., et al. (2018). Deep learning based classification system for identifying weeds using high-resolution UAV imagery. In *Science and information conference* (pp. 176–187). Springer.

Bechar, A., & Vigneault, C. (2016). Agricultural robots for field operations: Concepts and components. *Biosystems Engineering, 149*, 94–111.

Benson, T. (2019). *Robot apple pickers could change farming as we know it.* https://www.thedailybeast.com/georgia-gov-brian-kemp-risks-lives-to-reopen-his-state-amid-the-pandemic-just-not-his-familys?ref=scroll (Accessed 28 April 2020).

Berenstein, R., Ben Shahar, O., Shapiro, A., et al. (2010). Grape clusters and foliage detection algorithms for autonomous selective vineyard sprayer. *Intelligent Service Robotics, 3*(4), 233–243.

Birrell, S., Hughes, J., Cai, J. Y., et al. (2019). A field-tested robotic harvesting system for iceberg lettuce. *Journal of Field Robotics, 37*(2), 225–245.

Blanpied, G., Bramlage, W., Dewey, D., et al. (1978). In *A standardized method for collecting apple pressure test data.*

Bulanon, D. M., Kataoka, T., Okamoto, H., et al. (2004). Development of a real-time machine vision system for the apple harvesting robot. In *SICE 2004 annual conference* (Vol. 1, pp. 595–598). IEEE.

Chen, S. W., Shreyas, S. S., Dcunha, S., et al. (2017). Counting apples and oranges with deep learning: A data-driven approach. *IEEE Robotics and Automation Letters, 2*(2), 781–788.

Colin, L. (2019). *Berry picking robot hank could be perfectly timed for Brexit Britain*. https://agfundernews.com/berry-picking-robot-hank-could-be-perfectly-timed-for-brexit-britain.html. (Accessed 28 April 2020).

Culha, U., Nurzaman, S. G., Frank, C., et al. (2014). SVAS3: Strain vector aided sensorization of soft structures. *Sensors, 14*(7), 12748−12770.

Dalal, N., & Triggs, B. (2005). Histograms of oriented gradients for human detection. In *2005 IEEE computer society conference on computer vision and pattern recognition (CVPR'05)* (Vol. 1, pp. 886−893). IEEE.

Denny, O., Billingsley, J., & Reid, J. F. (2009). Agricultural robotics. *Journal of Field Robotics, 26*(6−7), 501−503.

Dong, C., Xu, Z., Dai, L., et al. (2019). Convolutional neural network-based approach for citrus diseases recognition. In *2019 IEEE intl conf on parallel & distributed processing with applications, big data & cloud computing, sustainable computing & communica- tions, social computing & networking (ISPA/BDCloud/SocialCom/SustainCom)* (pp. 1495−1499). IEEE.

Edan, Y., Han, S., & Kondo, N. (2009). Automation in agriculture. In *Springer handbook of automation* (pp. 1095−1128). Springer.

Edan, Y., & Miles, G. E. (1993). Design of an agricultural robot for harvesting melons. *Transactions of the ASAE, 36*(2), 593−603.

El Hariri, E., El-Bendary, N., Hassanien, A. E., et al. (2014). Automated ripeness assessment system of tomatoes using PCA and SVM techniques. In *Computer vision and image processing in intelligent syst. and multimedia technologies* (pp. 101−130). IGI global.

Escalante, H. J., Rodrıguez-Sánchez, S., Jiménez-Lizárraga, M., et al. (2019). Barley yield and fertilization analysis from UAV imagery: A deep learning approach. *International Journal of Remote Sensing, 40*(7), 2493−2516.

Evert van, F. K., Samsom, J., Polder, G., et al. (2011). A robot to detect and control broad-leaved dock (*Rumex obtusifolius* L.) in grassland. *Journal of Field Robotics, 28*(2), 264−277. issn: 1556-4959.

Eyarkai Nambi, V., Thangavel, K., & Jesudas, D. M. (2015). Scientific classification of ripening period and development of colour grade chart for Indian mangoes (*Mangifera indica* L.) using multivariate cluster analysis. *Scientia Horticulturae, 193*, 90−98.

Fairbank, W., Ede, L., Johnson, H., et al. (1987). Night picking. *California Agriculture, 41*(1), 13−16.

FB Klein, A. W., De Tejada, V. F., et al. (2019). In *Proof-of-concept modular robot platform for cauliflower harvesting*.

Foglia, M. M., & Reina, G. (2006). Agricultural robot for radicchio harvesting. *Journal of Field Robotics, 23*(6−7), 363−377.

Gebbers, R., & Adamchuk, V. I. (2010). Precision agriculture and food security. *Science, 327*(5967), 828−831.

Ge, Y., Xiong, Y., Tenorio, G. L., et al. (2019). Fruit localization and environment perception for strawberry harvesting robots. *IEEE Access, 7*, 147642−147652.

Gossett, S. (2019). *Farming agricultural robots*. https://builtin.com/robotics/farming-agricultural-robots. (Accessed 28 April 2020).

Habaragamuwa, H., Ogawa, Y., Suzuki, T., et al. (2018). Detecting greenhouse strawberries (mature and immature), using deep convolutional neural network. *Engineering in Agriculture, Environment and Food, 11*(3), 127−138.

Hajjaj, S. S. H., & Mohamed Sahari, K. S. (2016). Review of agriculture robotics: Practicality and feasibility. In *Robotics and intelligent sensors (IRIS), 2016 IEEE international symposium on* (pp. 194–198). IEEE.

Ha, J. G., Moon, H., Kwak, J. T., et al. (2017). Deep convolutional neural network for classifying Fusarium wilt of radish from unmanned aerial vehicles. *Journal of Applied Remote Sensing, 11*(4), 042621.

Han Lee, S., Chang, Y. L., Chan, C. S., et al. (2017). HGO-CNN: Hybrid generic-organ convolutional neural network for multi-organ plant classification. In *2017 IEEE international conference on image processing (ICIP)* (pp. 4462–4466). IEEE.

Han Lee, S., Chan, C. S., Paul, W., et al. (2015). Deep-plant: Plant identification with convolutional neural networks. In *2015 IEEE international conference on image processing (ICIP)* (pp. 452–456). IEEE.

Hanson, A. M. G. J., Joel, M. G., Joy, A., et al. (2017). Plant leaf disease detection using deep learning and convolutional neural network. In *International journal of engineering science*. Vol. 5324.

Harrell, R. C., Adsit, P. D., Munilla, R. D., et al. (1990). Robotic picking of citrus. *Robotica, 8*(4), 269–278.

Hayashi, S., Ganno, K., Ishii, Y., et al. (2002). Robotic harvesting system for eggplants. *Japan Agricultural Research Quar Terly, 36*(3), 163–168.

Hayashi, S., Shigematsu, K., Yamamoto, S., et al. (2010). Evaluation of a strawberry-harvesting robot in a field test. *Biosystems Engineering, 105*(2), 160–171.

Hemming, J., Bac, C. W., van Tuijl, B. A. J., et al. (2014). In *A robot for harvesting sweet-pepper in greenhouses.*

Hess, R. A. (2018). Human-in-the-loop control. In *Control system Applications* (pp. 327–334). CRC Press.

Hughes, J., Scimeca, L., Ifrim, I., et al. (2018). Achieving robotically peeled lettuce. In *IEEE robotics and automation letters.*

Hung, C., Nieto, J., Taylor, Z., et al. (2013). Orchard fruit segmentation using multi-spectral feature learning. In *2013 IEEE/RSJ international conference on intelligent robots and systems* (pp. 5314–5320). IEEE.

Hussain, A., Pu, H., & Sun, D.-W. (2017). Innovative nondestructive imaging techniques for ripening and maturity of fruits—A review of recent applications. In *Trends in food sci. & technology.*

Inkyu, S., Ge, Z., Dayoub, F., et al. (2016). Deepfruits: A fruit detection system using deep neural networks. *Sensors, 16*(8), 1222.

Jha, S. N., Narsaiah, K., Sharma, A. D., et al. (2010). Quality parameters of mango and potential of non-destructive techniques for their measurement—a review. *Journal of Food Science and Technology, 47*(1), 1–14.

Joseph, R., & Ali, F. (2018). Yolov3: An incremental improvement. In *arXiv preprint arXiv: 1804.02767*

Kader, A. A. (1997). Fruit maturity, ripening, and quality relationships. In *Int. Symp. Effect of pre-& postharvest factors in fruit storage* (pp. 203–208). Vol. 485.

Kamilaris, A., & Prenafeta-Boldú, F. X. (2018). Deep learning in agriculture: A survey. *Computers and Electronics in Agriculture, 147*, 70–90.

Kemp, C. C., Edsinger, A., & Torres-Jara, E. (2007). Challenges for robot manipulation in human environments [grand challenges of robotics]. *IEEE Robotics and Automation Magazine, 14*(1), 20–29.

Kiani, S., Azimifar, Z., & Kamgar, S. (2010). Wavelet-based crop detection and classification. In *Electrical engineering (ICEE), 2010 18th Iranian conference on* (pp. 587–591). IEEE.

Kondo, N., Yata, K., Iida, M., et al. (2010). Development of an end-effector for a tomato cluster harvesting robot. *Engineering in Agriculture, Environment and Food, 3.1*, 20–24.

Kusumam, K., Krajnik, T., Pearson, S., et al. (2016). In *Can you pick a broccoli? 3D-vision based detection and localisation of broccoli heads in the field.*

Li, W., Chen, P., Wang, B., et al. (2019). Automatic localization and count of agricultural crop pests based on an improved deep learning pipeline. *Scientific Reports, 9*(1), 1–11.

Li, L., Fan, Y., Huang, X., et al. (2016). Real-time UAV weed scout for selective weed control by adaptive robust control and machine learning algorithm. In *2016 ASABE annual international meeting. American society of agricultural and biological engineers* (p. 1).

Luo, L., Tang, Y., Zou, X., et al. (2016). Robust grape cluster detection in a vineyard by combining the AdaBoost framework and multiple color components. *Sensors, 16*(12), 2098.

McCool, C., Inkyu, S., Dayoub, F., et al. (2016). Visual detection of occluded crop: For automated harvesting. In *2016 IEEE international conference on robotics and automation (ICRA)* (pp. 2506–2512). IEEE.

Mehta, S. S., & Burks, T. F. (2014). Vision-based control of robotic manipulator for citrus harvesting. *Computers and Electronics in Agriculture, 102*, 146–158.

Mehta, S. S., MacKunis, W., & Burks, T. F. (2016). Robust visual servo control in the presence of fruit motion for robotic citrus harvesting. *Computers and Electronics in Agriculture, 123*, 362–375.

Moallem, P., Serajoddin, A., & Pourghassem, H. (2017). Computer vision-based apple grading for golden delicious apples based on surface features. *Information Processing in Agriculture 4, 1*, 33–40.

Mohamed, K., Hafiane, A., & Canals, R. (2018). Deep leaning approach with colorimetric spaces and vegetation indices for vine diseases detection in UAV images. *Computers and Electronics in Agriculture, 155*, 237–243.

Monta, M., Kondo, N., & Shibano, Y. (1995). Agricultural robot in grape production system. In *Robotics and automation, 1995. Proceedings., 1995 IEEE international conference on* (Vol. 3, pp. 2504–2509). IEEE.

Monta, M., Kondo, N., & Ting, K. C. (1998). End-effectors for tomato harvesting robot. In *Artificial intelligence for biology and agriculture* (pp. 1–25). Springer.

Mustafa, N. B. A., Ahmad Fuad, N., Ahmed, S. K., et al. (2008). Image processing of an agriculture produce: Determination of size and ripeness of a banana. In *Inf. Technology, 2008. ITSim 2008. Int. Symp. On* (Vol. 1, pp. 1–7). IEEE.

Nieuwenhuizen, A. T., Hofstee, J. W., & van Henten, E. J. (2010). Performance evaluation of an automated detection and control system for volunteer potatoes in sugar beet fields. *Biosystems Engineering, 107*(1), 46–53.

Ng, A., et al. (2011). Sparse autoencoder. *CS294A Lecture Notes, 72*(2011), 1–19.

Nicolas, H., Maiolino, P., Iida, F., et al. (2018). A variable stiffness robotic probe for soft tissue palpation. *IEEE Transactions on Robotics, 3.2*, 1168–1175.

Ojala, T., Pietikainen, M., & Maenpaa, T. (2002). Multiresolution grayscale and rotation invariant texture classification with local binary pat- terns. *IEEE Transactions on Pattern Analysis and Machine Intelligence, 24*(7), 971–987.

Pfeifer, R., Iida, F., & Lungarella, M. (2014). Cognition from the bottom up: On biological inspiration, body morphology, and soft materials. *Trends in Cognitive Sciences, 18*(8), 404–413.

Ceres, R., Pons, J. L., Jimenez, A. R., et al. (1998). Design and implementation of an aided fruit-harvesting robot (Agribot). In *Industrial robot: An international journal*.

Rebetez, J., Satizábal, H. F., Mota, M., et al. (2016). Augmenting a convolutional neural network with local histograms-A case study in crop classification from high-resolution UAV imagery. In *Esann*.

Reddy, N. V., Vishnu Vardhan Reddy, A. V., Pranavadithya, S., et al. (2016). A critical review on agricultural robots. *International Journal of Mechanical Engineering and Technology, 7*, 4.

Sams, C. E. (1999). Preharvest factors affecting postharvest texture. *Posthar Vest Bio. and Technology, 15*(3), 249–254.

Scarfe, A. J., Flemmer, R. C., Bakker, H. H., et al. (2009). Development of an autonomous kiwifruit picking robot. In *2009 4th international conference on autonomous robots and agents. IEEE* (pp. 380–384).

Scimeca, L., Maiolino, P., Cardin-Catalan, D., et al. (2019). Non-destructive robotic assessment of mango ripeness via multi-point soft haptics. In *2019 international conference on robotics and automation (ICRA)* (pp. 1821–1826). IEEE.

Scimeca, L., Maiolino, P., & Iida, F. (2018). Soft morphological processing of tactile stimuli for autonomous category formation. In *2018 IEEE international conference on soft robotics (RoboSoft)*. IEEE.

Shah Rizam, M. S. B., Yasmin, A. R. F., Ihsan, M. Y. A., et al. (2009). Non- destructive watermelon ripeness determination using image processing and artificial neural network (ANN). *World Academy of Science, Engineering and Technology, 38*, 542–546.

Siegner, C. (2019). *Tomato harvesting robot grows $1.5M in seed funding*. https://www.fooddive.com/news/tomato-harvesting-robot-grows-15m-in-seed-funding/560093/. (Accessed 28 April 2020).

Slack, G., & Hand, D. W. (1983). The effect of day and night temperatures on the growth, development and yield of glasshouse cucumbers. *Journal of Horticultural Science, 58*(4), 567–573.

Slaughter, D. C. (2009). *Nondestructive maturity assessment methods for mango* (pp. 1–18). Davis: University of California.

Staff, S. P. W. (2019). *Robotic strawberry harvester one step closer to reality in Florida fields*. https://southeastproduceweekly.com/2019/04/16/robotic-strawberry-harvester-one-step-closer-to-reality-in-florida-fields/. (Accessed 28 April 2020).

Tanigaki, K., Fujiura, T., Akase, A., et al. (2008). Cherry-harvesting robot. *Computers and Electronics in Agriculture, 63*(1), 65–72.

Van Henten, E. J., Hemming, J., Van Tuijl, B. A. J., et al. (2002). An autonomous robot for harvesting cucumbers in greenhouses. *Autonomous Robots, 13*(3), 241–258.

Van Henten, E. J., Van Tuijl, B. A. J., Hoogakker, G. J., et al. (2006). An autonomous robot for de-leafing cucumber plants grown in a high-wire cultivation system. *Biosystems Engineering, 94*(3), 317–323.

Wang, X., & Cheng, C. (2015). Weed seeds classification based on PCANet deep learning baseline. In *2015 asia-Pacific signal and information processing association annual summit and conference (APSIPA)* (pp. 408–415). IEEE.

Williams, A. (2019). *Fieldwork Robotics completes initial field trials of raspberry harvesting robot system*. https://phys.org/news/2019-05-fieldwork-robotics-field-trials-raspberry.html. (Accessed 28 April 2020).

Youngman, A. (2017). *Robot trials aim to cut strawberry harvesting time and control costs.* https://www.producebusinessuk.com/purchasing/stories/2017/02/07/robot-trials-aim-to-cut-strawberry-harvesting-time-and-control-costs. (Accessed 28 April 2020).

Yue, Yi., Cheng, Xi., Zhang, Di., et al. (2018). Deep recursive super resolution network with Laplacian Pyramid for better agricultural pest surveillance and detection. *Computers and Electronics in Agriculture, 150*, 26—32.

Zhao, D.-A., Lv, J., Wei, Ji., et al. (2011). Design and control of an apple harvesting robot. *Biosystems Engineering, 107*(2), 112—122.

Capturing agricultural data using AgriRover for smart farming

Xiu-Tian Yan[1]**, Cong Niu**[1]**, Youhua Li**[1]**, Willie Thomson**[2]**, Dave Ross**[2]**, Ian Cox**[3]

[1]*Robotics and Autonomous Systems Group, The University of Strathclyde, Glasgow, United Kingdom;* [2]*Agri-EPI Centre, Northern Agri-Tech Innovation Hub, Midlothian, United Kingdom;* [3]*Innovate UK, Swindon, United Kingdom*

Introduction

Agricultural industry is experiencing a rapid and revolutionary change empowered by the use of abundant data generated by new sensory devices. These range from mobile and small devices such as hand-held soil quality measurement instruments, to large equipment such as satellites which are positioned to generate large amount of imagery data as well as spectral data of crops. Recent technology development in earth observation, enabled by bespoke agricultural satellites such as NASA's Soil Moisture Active Passive satellite, is transforming the agricultural production from a physical action-intensive production process to data-guided food production process. This shift provides an opportunity for farmers to be able to tailor their food production process to be smarter, more efficient and more sustainable.

For example, the Soil Moisture Active Passive satellite collects data about the global soil moisture and this critical information can be used to guide farmers to irrigate intelligently with optimised provision of water to crops. Similarly, in the United Kingdom, Sentinel 1 and 2 were designed and launched to provide data to cover crop products, which are freely available. Products derived from these sources range from identifying and monitoring pests and disease, monitoring soil fertility and water status, providing guidance on the amount of crop nutrition application, and to helping to plan how and what amount of water should be used for irrigation. They can also be implemented to monitor and inform farmers on the best time for harvesting and estimating and predicting yields.

To generalise the process of this revolutionary transformation, it is useful to draw a generic diagram to show the process of data capturing and collection found in agricultural industry. Fig. 5.1 shows the architecture of generic and typical data collection process, typically found in an agricultural application based on several projects that authors are involved in.

Fig. 5.1 also shows a flow block diagram of modules and sub-modules for data capturing processing and execution. The data capturing box contains the function modules of robotic platform where first the data are captured with sensors deployed

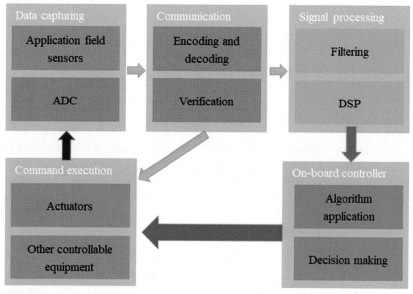

FIGURE 5.1

An architecture and modules of data capturing, processing and execution.

in an application field and then digitised using typically a device called Analog to Digital Converter. The original data can also come from Command execution box generated by devices such as actuators to monitoring each of such devices and other controllable equipment from which data can also be generated. The converted digital data using the modules in the communication box go through further processes, which include coding and signal integrity verification before the digital signal/data is send to the signal processing module. In the signal processing module, the digital signals and data are filtered and processed using digital signal processing (DSP). After the DSP in the on-board controller module, the on-board CPU uses the data for decision-making by deploying appropriate algorithms. Finally, the commands based on the decisions are sent to the execution modules where the actuators of the robotic platform such as wheels and arms are activated to move or enact certain equipment such as lights and cameras for correct manipulation. This completes the full cycle of data capture and utilisation and this iterative process continues till the next state change within the system.

SmartFarm: a holistic philosophy for UK—China Agritech collaboration philosophy

As a specific example, this section introduces the work undertaken by the authors in an international collaboration project entitled SmartFarm constructed by two groups of researchers from the United Kingdom and China. The SmartFarm Philosophy is a new holistic and systematic approach to addressing the global food production challenge by gaining an insightful understanding of food production systems and developing suitable technologies to enable the knowledge-based interventions. The SmartFarm Philosophy is designed to significantly improve the conversion of inputs such as nutrients and energy into high-quality food in a sustainable manner. The key to this is the holistic understanding and creation of an efficient food production system of distributed crop and livestock systems. This approach will create a greater understanding of whole food supply chain efficiency and better integration of these systems, generate long-term positive environmental impact and deliver greater food security. The Philosophy brings long-term purpose, and environmental stewardship to the entire agri-food system. The SmartFarm can hence potentially transform food production in the future, enabled by smart technologies for increased yields and improved food quality whilst aiming to reduce the usage of chemicals and other inputs.

The SmartFarm project is an important part of the UK China Agritech Flagship Challenge and aims to pilot the development of a world leading solution to food security challenge. This will be achieved by smart technologies developed through UK—China collaborative innovation. It builds upon the UK's capacity to develop and integrate smart sensory and robotic technologies for the purpose of monitoring, intelligent decision support, action planning and timely and effective (targeted) intervention.

Benefits of adopting the SmartFarm philosophy

This new paradigm shift is aligned well with the recent development trends and desire in global energy production and consumption to reduce the impact of fossil fuel on climate change. The proposed SmartFarm approach will address the overall challenge in increasing the efficiency of energy absorption and conversion from solar source to food human being consumed. Specifically, the SmartFarm philosophy promotes a suite of concepts in increasing energy conversion by smartly monitoring the growth of both crops and livestock, and timing interventions intelligently. It is possible to identify the following specific benefits:

- A new understanding of energy and nutrient flow and utilisation through the primary agri-food system will be established.
- Efficient and targeted support for key science initiatives can be initiated.
- A new mechanism can be developed to support the development, and implementation of agricultural and environmental policy priorities.
- More efficient and sustainable food production with improved environmental stewardship can be enabled.
- Creation of a modern, technically advanced and dynamic SmartFarm equipment industry and supply chains can be forged.

- Clear monitoring of progress and impact of investment is embedded through the large data generated from the SmartFarm approach.
- New data sets generated can stimulate significant innovation by developing solutions for new challenges and a healthy, safe and tasty food diet.
- The proposed SmartFarm also has a huge potential of creating a legacy exemplar for successful international cooperation.

There are also a number of challenges to a successful SmartFarm which have been identified from a technical and international collaborative perspective. These include the following:

- Agreement on accessing and sharing data for greater scientific and business innovation in the context of international cooperation for better human lives;
- Adequate allocation of funding to support accelerated and collaborative innovation by combining mutual scientific and innovative expertise and skills;
- Identification of risks and mitigation measures to ensure sustained successful collaboration for secure and environmentally sustainable food production.

Design methodology for a mobile data collection platform

As an example of data collection, this section describes a design methodology used in designing a mobile data collection platform called AgriRover. This is one of many possible examples newly designed to increase the amount of data and frequencies of data collection for soil quality monitoring. This chapter focuses on development of such a design methodology and then covers the energy aspect of AgriRover, both in its navigation and in data collection.

Based on the development of a design methodology (Yan & Zante, 2010) and the research (Hou et al., 2019) and review work (Rubio et al., 2019) and (Kashiri et al., 2018), it is clear that mobile robotic systems require energy-efficient path planning in order to traverse its fields effectively. This energy efficiency requirement is especially true when the mobile robots are deployed in an unknown environment where effective navigation, enabled by perception capability of the environment, localisation and cognition to make decisions, is essential for achieving its goals, as reviewed by Rubio et al. (2019).

Before generating an algorithm for energy-efficient path planning, it is necessary to investigate generic design methodology that considers all important aspects of the system (including environmental factors) so that the proposed work will be generic and applicable in more applications. In this chapter, a purpose-generated design methodology is presented for path planning of mobile robotics. An energy optimisation algorithm can then be derived for detailed modelling and path optimisation to improve energy efficiency and long-distance operations. From literature review, a mechatronic V model is considered, and the algorithm design is modelled and developed based on the mechatronic design model reported by Yan and Zante (2010).

The proposed design process model consists of three pillars, namely information repository pillar, design process pillar and energy pillar. The information repository pillar is concerned with the representation of the design information, which grows when more design decisions are made and hence more design parameter values are committed. The design process pillar represents the conventional process of designing mechatronic systems such as the mobile AgriRover. These processes flow from top to bottom of Fig. 5.2 and help a designer to progress to a satisfactory design. Building on these, work to extend design methodology is proposed, by introducing a new pillar which focuses on the consideration of energy conversion and external impact factors on field robots. Specifically, the methodology now includes the consideration of the dynamically changing environment and its impact on the energy consumption of a mobile robotic system. These factors include the elevation of the terrain, the surface unevenness of the terrain and the hardness of the surfaces that

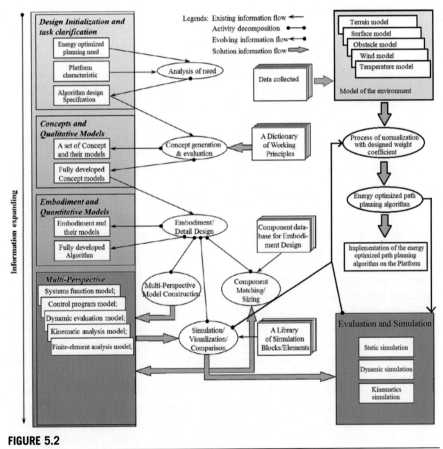

FIGURE 5.2

A three-pillar mechatronic design methodology focusing on energy design for mobile platforms.

a mobile AgriRover needs to travel. With these considerations in the design process model, it is anticipated that a full and comprehensive consideration for a systematic design can be achieved to ensure a mobile robot is investigated to be energy efficient by design.

In the information repository pillar on the left of Fig. 5.2, the design information of the mechatronic system design process is displayed, and it gradually expands as the system design decisions are made, until the relationships between each interconnecting part are clarified and decided. This evolving process is shown on the left with the evolving information flow arrow. This evolving information can be categorised into four groups of information with different levels of detail from the top to the bottom of the pillar. These include information on design initialisation and task clarification, which can be further divided into the type that describes the system level needs of customers, specifications through analysis and also the market intelligence for commercialising a mobile robot for a particular market. The information becomes richer as shown in Fig. 5.2 by the increasingly darkening colour on the robot design specification. This evolving information flow is enriched from the existing design information to the next stage. At the end, there is a design solution in the form of a mixture of the statements of the customers' needs, market research results, conceptual design solution models, embodiment design and importantly, a set of design models from multiple perspectives, namely control design model. All these help to make decisions on the robotic design.

In the middle pillar, from the top to the bottom, the second pillar in Fig. 5.2 shows a mechatronic system design process combined with activities that generate a final design solution. It involves both concepts generation and qualitative modelling. In this step, concepts are generated and evaluated according to specific working principles, with the existing information of design specifications produced in step one. After that, a fully developed concept model is created for further evaluation.

The information from a concept generated and evaluated is passed to the third step, which is embodiment design and quantitative modelling (for evaluation). Embodiment and detail design are first undertaken so that all components are specified. Then the evolving information is incorporated into the embodiment models, to generate fully developed solution models. These models can then be simulated and evaluated for quantitative evaluation.

Finally, the last part of the mechatronic system design process in the middle pillar is about multi-perspective modelling and simulation, in which a mobile robot solution is fully modelled and simulated from several perspectives as illustrated in the last of the information repository pillars. All these multi-perspective models provide a full representation of a mobile system for intensive evaluation. There are more intertwined relationships and trade-off required and considered at this stage for an optimal solution as shown in Fig. 5.1. In this step, many models are designed as shown in the button on the left of the figure. Building on the library of simulation models and simulation and visualisation, a mechatronic system can be fully generated, evaluated and validated in this virtual modelling world. This forms the basis of the mechatronic system design process model reported by Yan and Zante (2010).

From the energy optimisation point of view, the design process model shown in Fig. 5.2 addresses the following aspects: the environment, path planning needs and the targeted terrain.

Terrain surface modelling is concerned with representation of surface evenness of a field and shows the difference between normally frictionless surface and realistic terrain surface on which a mobile platform transverse. On top left, the planning need is to be formulated, including requirements such as the shortest path and lowest energy cost, which should be clarified. Moving down to the second part, the platform characteristics need to be considered, different path planning method to be utilised. After the conceptual design in Fig. 5.2, the remainder of the design stages remain the same for the concept and qualitative design, and the embodiment and quantitative design.

Lastly, as this is an energy-optimised planning design concept, the number of modelling types is less compared to the universal design concept.

On the right is the design of path planning algorithm. From the top, the data collected with various methods (such as data collected by the on-board and off-board sensor, weather station, satellite) will be modelled according to the requirement of the application, which could include air-based/ground-based/water-based use cases. After the modelling of the environment, there is a processing algorithm for each of the environmental models that gives the different priorities according to the objectives of the specific platform, which could include maximum distance travelled, full coverage, exploration and more. According to these objectives, each of the environmental models is processed and selected accordingly. With selected inputs, the designed energy-optimised algorithm will generate a path for the specific platform independently, which is flexible and expendable. The result of the path generated is then simulated, and the parameters of the planning algorithm may be fine-tuned for the improvement of the performance of the energy-optimised path planning algorithm. When the performance is determined as satisfactory, the algorithm may be implemented to the individual platform, and the field test can be performed.

AgriRover design

Following the above design process and taking the key considerations discussed, an AgriRover (Yan et al., 2020, pp. 55–73) is designed as a multi-purpose autonomous mobile platform specifically for agricultural applications. It has two main functions: monitoring soil condition by providing instant measurement of soil fertility and selectively harvesting for targeted fruits for the overall development of smart farming. Cost minimisation is a key driver of the design, and this leads to the considerations of operational cost, and energy usage from on-board power source, in order to maximise the economic benefits to farmers. The key energy components of AgriRover utilise a set of rechargeable lithium batteries as the energy source, and they drive electrical actuators and motors as the mobility system. The AgriRover is designed with the focus on eco-friendly and sustainable development, as it is important both environmentally and financially in long term. One of the design perspectives is to make the AgriRover energy efficient during its operation (Fig. 5.3). This can be achieved by following an energy efficient pathway using an energy model detailed in Section Design methodology for a mobile data collection platformn (Fig. 5.3).

Mechatronic approach with a focus on energy consumption of the systems

The sustainable development of food production is becoming an important requirement. Without the consideration of sustainability it will become a 'contemporary issue' as Klarin introduced (Lakshmanan et al., 2020). Even though the coal consumption for generating electricity has decreased since 2003 from 38.4% to 22.4% of electricity generated in 2019, the electricity generated from fossil fuel is still 53.8% of all electricity generated according to the International Energy Organization (IEA) in 2020. For mechatronic and robotic system design, being electricity efficient is becoming more important for sustainability. An energy-optimised path planning method is a such solution for better sustainable solution, achieving potentially up to 20% energy saving in a rich terrain environment.

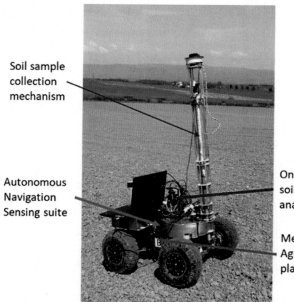

FIGURE 5.3

AgriRover prototype performing soil sampling and monitoring.

Overview of navigation algorithms and autonomous field vehicle platforms

Overview of navigation systems

As one of the most important parts of any mobile robotic platform, the navigation sub-system needs to be designed to cope with many unknown conditions, in a relatively unstructured working environment (represented by many different types of farm fields). Otherwise, the capability of the robotic platform is not robust and its normal functional behaviour will suffer in more challenging farm fields.

Fundamentally, there are five mainstream technologies or approaches which are available to construct a navigation system, which include GPS, LIDAR, inertia measurement unit (IMU), vision-based systems and ultrasonic sensor-based systems such as the ones used in parking sensors. Each technology has its own advantages and limitations, where in some applications, one is more preferred than another. In most of the applications, multiple sensory systems deploy a suite of such sensors with a combination of both different type and different numbers of the same type sensors in order to achieve a desirably accurate tracking and detection of potential obstacles. Although the hardware of the navigation system is important, the navigation algorithms play a more significant part in the final performance of the navigation system and next section introduces main stream navigation algorithms by comparing them for more energy-efficient navigation for data collection.

A review of navigation algorithm

For the purpose of an energy-efficient path planning algorithm, a comprehensive review of published navigation algorithms was undertaken and most of the mainstream path planning algorithms compare to show characteristics that suit the different navigation requirements. The results are shown in Table 5.1.

Table 5.1 compares most of the mainstream path planning algorithms for mobile robotic platforms from five perspectives in order to assess their strengths and weakness. Different path planning algorithms have different characteristics on applicability and the global objectives of path planning. Some path planning algorithms are more suitable for specific applications where coverage is more important than point to point exploration or trajectory planning. All the path planning algorithms are rated with a number of stars in a scale of 1−5 starting from one of the five criteria shown in the first row of Table 5.1.

First criterion is about the suitability of modelling different terrain and ground roughness. This is evaluated based on the path planning principles each of the algorithms has developed. For example, algorithms such as RRT or A* plan a path according to a dynamic matrix which changes every step as the waypoint is generated. Therefore, the fixed terrain and ground roughness is not suitable, as these roughness characteristics change and need to match the dynamic characteristics of the evaluation matrix, which requires extra complexity and computation power. It

Table 5.1 Comparison and characteristics of algorithms for energy optimised path planning.

Name	Terrain roughness modelling suitability	Computational power needed	Extendibility	Flexibility	Robustness
A*, D* algorithm	★	★	★★★	★★	★★★
RRT	★	★★★	★★★★	★★★★	★★★★
Pure pursuit	★	★	★★★	★	★★★★★
Artificial potential field	★★★	★	★★★	★★★	★★
Particle swarm	★★	★★★	★★	★	★★
Ant colony	★★★	★★★	★★	★★	★★
Genetic algorithm	★★★	★★★	★	★★★	★
Neural network	★	★★★★		★	★
Machine learning	★★	★★★★	★	★	
Model predictive	★	★★★		★	★★★

is concluded that RRT and A* navigation algorithms are less suitable for terrain and ground roughness modelling.

Secondly, computational power is an important requirement which is measured through the computational resources required, when the path planning algorithm is running and when the map is dynamically being expanded. Some path planning algorithms such as Machine Learning or Neural networks will require new training when the map is changed, which requires high computational resources. When the map size is expanded, the computational resources cost for the training will increase nonlinearly and significantly. Fewer stars mean the computational power requirement is less and the scaling of the map size has less effect in coping with more challenging terrains.

Thirdly, the extendibility is a measure of the possibility of extending the capability of an algorithm to other robotic systems, as needed. For example, more environmental elements such as the properties of the surface need to be considered for larger vehicles in order to be sure that the surface is not too soft and is suitable to travel. Similarly, these algorithms should also be possible to be extended for crop planting robots, irrigation robots and pesticide applying robots. Adding more environmental considerations for a navigation planning algorithm such as Genetic Algorithm will require a complete redesign and reconfiguration of the algorithm, which would require similar amount of manpower compared to designing a completely new algorithm. This will waste time and money and should be avoided. Therefore, more stars in Table 5.1 are preferred as they indicate a better suitability for extending the functions of the algorithm for new application requirements.

Flexibility is defined as the suitability of transferring and redeploying algorithms from an existing platform to a new platform on which they can work. For example, the fully developed energy-optimised path planning algorithm for a mobile robotic platform has the possibility of being re-useable onto a new unmanned aerial vehicle. Flexibility assesses how much the original codes need to be altered and modified for the new application. The more stars an algorithm is rated, the better its flexibility is. Flexibility also refers to the possibility of changing the accuracy and performance of the algorithm as a trade-off with regard to the computational power required. For example, varying the size of path finding matrix can improve the accuracy, whereas reducing the frequency of updating the matrix can lower the computational power requirement. An algorithm should have the capability of changing its performance by increasing the update frequency or the size of path finding matrix in order to enable quicker and better real-time response of the system for better dynamic responsiveness.

Finally, the robustness is an evaluation based on the performance stability of the algorithm when it is being used during the real-world applications, where there are uncertain considerations that are imposed upon the algorithm. It indicates whether or not an algorithm would have been affected significantly. A high degree of robustness of an algorithm suggests it should reliably cope with unknown and uncertain conditions by producing robust plan. The algorithms that are more complex have a tendency of being less stable under the test of uncertain environments.

In conclusion, designing an energy optimised path planning algorithm for the application on an agriculture autonomous mobile robotic platform requires the consideration of five criteria shown in Table 5.1. Artificial Potential Field algorithm is the most suitable for this application on balance. First, as the potential field can be superimposed and combined, the extensibility and terrain modelling suitability scores high with three stars, indicating it is suitable to model unknown terrains and it is also achievable to extend from algorithms developed for other system to a new one. Secondly, for the Artificial Potential algorithm, the pathfinding process is based on the attraction and repulsion of an artificial field. The size of search field can be changeable which provides more flexibility in planning accuracy and adaptable to a new terrain. These requirements point to the Artificial Potential algorithm more suitable for online and on-board real-time path planning as it considers gravity potentials. In conclusion, the Artificial Potential Field is the most suitable algorithm for energy-optimised path planning that only requires minimum amount of computational power and offers great expandability.

Energy-focused mechatronic modelling

For energy-efficient mobile AgriRover design, a list of considerations has been compiled based on the literature review and some field studies. These can then be used to help to build a comprehensive representation of energy models of a mobile AgriRover design.

A total energy model is developed for total energy P_T, considering all critical energy used in a typical rover system and this can be derived by the following:

$$P_T \int_{t_1}^{t_2} \left(P_m + P_s + P_f + P_h + P_e \right) \cdot dt$$

where P_m is the summation of the power on each wheel and at any given time, $P_m = F \cdot V_1$. The P_s is the summation of the power consumed by the steering system, calculated $P_s = \tau \cdot \omega$. To overcome the friction between the rover and contact surface, P_f is introduced. The power due to this heat loss by the propulsion driving system can be defined by P_h. P_e is the power required for the on-board electronic equipment. With this model, the total energy consumed is represented to provide a measure for path planning. Fig. 5.4 shows an example of a comparison of energy consumptions between a traditional straight line path plan and an energy optioned path plan. The top bar shows the results from simulation and the bottom bar shows the results from field study of the same terrain.

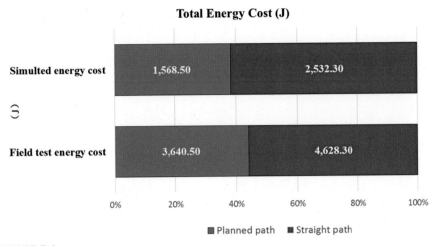

FIGURE 5.4

AgriRover prototype performing soil sampling and monitoring.

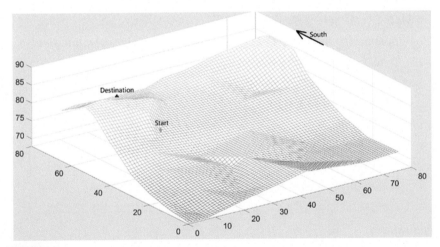

FIGURE 5.5

An example terrain map created from a chosen field for data collection.

UK–China SmartFarm data flow diagram
Path planning

In order to capture essential soil data of a targeted field, it is necessary to plan the soil data collection path. A terrain map of the field can be constructed by accessing the field information from online data service provider such as Digimap by EDINA, a web mapping and online data delivery service developed by the EDINA national data centre for UK academia. An example terrain map is shown in Fig. 5.5 for a field trial in the United Kingdom. The map has a size of 66 by 76 data points, covering a field of 330 by 380 m.

An energy efficiency path plan generation algorithm has been developed and for any given terrain digital map; it is applied to generate an optimal path plan. This basically applies the following pseudo-code to search for an optimal energy efficient path.

Find the starting position
Pseudocode:

```
integer n_x, n_y, i_field_x, i_field_y, j,m, k, l_destin
double potential[1..n, 1..m], merged_potential[1..n, 1..m]
double position_x[1..n],  position_y[1..m],  way_point[1..nwp_x, 1..nwp_y]
i_field_x ← 1
i_field_y ← 1
while i_field_y ← 1 <=n_y do
   while i_field_y ← 1 <=n_y do
      potential[i_field_x, i_field_y] ← compute_height_potential_field
      i_field_y ← i_field_y+ 1
   endwhile
   i_field_x ← i_field_x+ 1
endwhile
j ← 1
while j<=k do
      potential[j]  ← compute_distance
      i ← i+ 1
endwhile
merged_potential[1..n, 1..m]  ← potential[1..n, 1..m]
j ← 1
while j<=l_destin do
      way_point[wp_x, wp_y]  ← compute_way_point[j,wp_x, wp_y]
      j ← j+ 1
endwhile
```

After applying the energy efficiency path plan generation algorithm, an energy optimisation has been performed and optimised energy path is generated. The path line is shown Fig. 5.6. As the figure shows, the height has increased from 79 to 82.2 m and then the height changes remain between 81.43 and 82.25 m. The total cost of energy of 972.65 J can be observed in this case.

Power data capture in field studies

Using the above system design for data capture, many field trials have been undertaken both in the United Kingdom and in Beijing, China. Multiple data packages have been captured and these include soil data as well as the energy consumption data by the AgriRover. These data are stored on board for later retrieval and processing.

All captured data are processed and one example is to calculate the power in watts by multiplying the voltage to the duty cycle and then the current which gives the power in watts. Four sample data sets are shown in Fig. 5.7 for monitoring the power consumption of the driving sub-system as well as the steering sub-system. Fig. 5.7 (1) shows the power needed for the steering motion of the AgriRover. Fig. 5.7 (2) indicates the driving power required to move the rover along an optimised path based on potential energy algorithms developed by the authors. Fig. 5.7 (3) similarly shows the power needed for the steering motion of the rover from four wheels, and in this case, it shows the power needed for the steering the rover when following a straight path, whereas Fig. 5.7 (4) is the result of the power needed for the driving motion of the rover going in a straight line.

From these test results, the following analyses are undertaken. The summation of power at each instant in Fig. 5.7 (2) is lower than that of Fig. 5.7 (4), which suggests the driving system of the AgriRover experiences less change of power consumption than that on a straight line. The case in Fig. 5.7 (2) also has lower overall peak power required. Observation shows when driving uphill with a higher incline angle, sometimes only three of the four wheels have contact with ground, resulting in increased

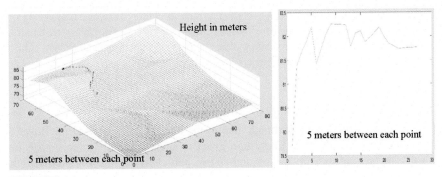

FIGURE 5.6

An optimised path line is generated and can be visualised from start to destination.

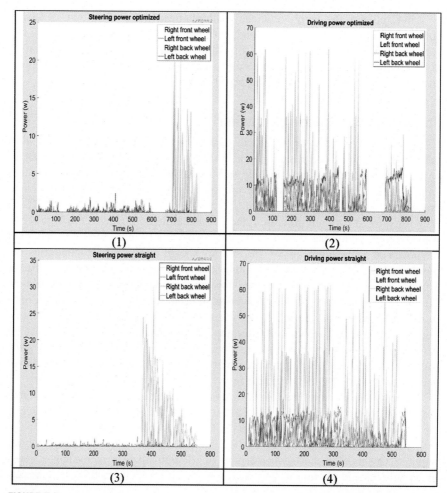

FIGURE 5.7

The power outputs from each driving and steering module captured during a field test. (1) power on steering sub-system for an optimised planned path (2) driving sub-system for an optimised planned path, (3) four steering power outputs during a traditional straight path plan and (4) driving power for the straight path.

power requirements on a single wheel to maintain momentum. For Fig. 5.7 (2), which shows energy optimised path, due to the lower incline needed for Rover to overcome, there is also less likely in a given moment that only three out of four wheels are in contact with the ground at this particular testing. Furthermore, the time required to finish is longer under condition in Fig. 5.7 (2) (the energy optimised planning) compared to that in Fig. 5.7 (4) (the straight path).

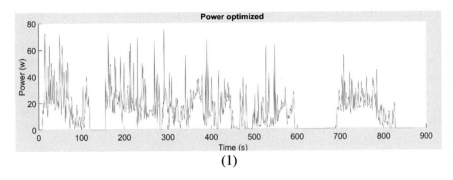

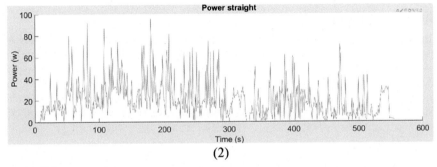

FIGURE 5.8

Total power at a given time of the AgriRover for an optimised path in comparison with that for a straight path.

Adding the above power on all driving and steering motors provides the total power consumed and is plotted in Fig. 5.8 (1) which shows the output power at a given time for the planned path. It can be seen from this chart that the maximum power at a given time is less than 80 W at any given moment with the standard deviation of 14.38.

Fig. 5.8 (2) is the output power at a given time when the AgriRover was running on the straight path normally adopted by most planners. The figure shows for the straight path, the maximum power at a given time is over 80 W at multiple times. It has a standard deviation of 17.27, suggesting this set of output power at a given time has higher variation or varied loading on the driving and steering systems, which can lead to accelerated wear of these bearings and motors.

Conclusions and future work

A comprehensive approach has been described for a new agricultural robot called AgriRover to navigate in a farm field. This chapter describes an energy-efficient path planning for such a rover which could also be extended to other mobile platforms. This development has been implemented in an overall smart farming project aiming to increase the smartness level in farming significantly by deploying advanced agricultural technologies such as robotics and other data capturing systems (such as satellites and hand-held devices). A data capture and connection architecture has been proposed in order to enable an efficient data capturing and processing so that future smart farming is empowered by rich data for timely and smart farming decision. With this foundation work, the system demonstrated its potentials and future work is required to further optimise its capability in data capture and improve the overall robustness of the system.

Acknowledgments

The authors of the chapter would like to thank Innovate UK and UK Space Agency for providing funding to develop an AgriRover for trial under SmartFarm project. The authors would also like to acknowledge the significant contribution towards AgriRover prototyping by colleagues from RAL Space. The authors would also like to thank and acknowledge the provision of NERCITA farm in Beijing for the testing and construction of the UK−China SmartFarm and the support of China Rural Technology Development Centre (CRTDC) for the SmartFarm construction.

References

Hou, L., Zhang, L., & Kim, J. (2019). Energy modeling and power measurement for mobile robots. *Energies, 12*, 27. https://doi.org/10.3390/en12010027
Kashiri, N., Abate, A., Abram, S. J., Albu-Schaffer, A., Clary, P. J., Daley, M., Faraji, S., Furnemont, R., Garabini, M., Geyer, H., Grabowski, A. M., Hurst, J., Malzahn, J., Mathijssen, G., Remy, D., Roozing, W., Shahbazi, M., Simha, S. N., Song, J.-B., ... Tsagarakis, N. (2018). An overview on principles for energy efficient robot locomotion. *Frontiers Robotics AI, 5*, 129. https://doi.org/10.3389/frobt.2018.00129
Lakshmanan, A. K., et al. (2020). Complete coverage path planning using reinforcement learning for tetromino based cleaning and maintenance robot. *Automation in Construction, 112*, 103078.
Rubio, F., Valero, F., & Llopis-Albert, C. (March-April). A review of mobile robots: Concepts, methods, theoretical framework, and applications. *International Journal of Advanced Robotic Systems*, 1−22. https://doi.org/10.1177/1729881419839596
Yan, X.-T., Bianco, A., Niu, C., Palazzetti, R., Henry, G., Li, Y., Tubby, W., Kisdi, A., Irshad, R., Sanders, S., Scott, R., et al. (2020). The AgriRover: A reinvented mechatronic platform from space robotics for precision farming. In *Reinventing mechatronics*. Springer.
Yan, X.-T., & Zante, R. (2010). A mechatronic design process and its application. In *Mechatronics in action* (pp. 55−70). Springer.

Health and welfare monitoring of dairy cows

6

Toby Trevor Fury Mottram[1], Ingrid den Uijl[2]

[1]Digital Agritech Ltd, Kirkcaldy, United Kingdom; [2]Cowmanager, Harmelen, Netherlands

Introduction

Historically, herd health management relied on human memory and visual observation. Data, when it was available at all, was qualitative and recorded in hand written notes in a diary. The first affordable farm computers with useable software began to appear in the early 1980s and the first electronic milk recording systems began to appear after about 1985. These records were mainly used for changing feeds and selecting cows for breeding, evaluating events after the fact. True digitisation became possible as sensors became directly connected to computers. The masters project of the author in 1987 fitted strain gauges to a milk recorder jar. Sending the data about milk yield via dial up modem to a printer at Silsoe College 100 miles away was a real technical achievement. Since then, new techniques of monitoring animal health and welfare automatically and remotely have become commonplace. On farms, sensors directly connected through a network, software models, databases, farm computers, internet connection and cloud-based analysis are normal, although by no means widely adopted. Currently, with connection to cloud databases, very complex models can be computed to detect, predict or prevent adverse events from happening and automate routines and workflows.

Dairy cows are valuable animals that have the potential to live long productive lives. The principal reason for these animals failing to achieve their lifetime potential is poor fertility management, calving problems, mastitis, metabolic conditions and lameness (Mottram, 1997). Although automated monitoring systems have a cost of installation and maintenance, these can be justified by a number of factors that improve herd performance and animal welfare.

The speed of development of new systems is very rapid so the focus of this chapter is on the general principles as to how the systems work and the many approaches in research. Many of the problems identified in earlier papers can be overcome by software processing and integration with other sensor and historic data. Presentation of farm data via internet is still a challenge, in part because of poor internet connections.

This chapter focuses on the technologies currently available and in development for monitoring dairy cow health to minimise animal suffering and losses through disease. The freedom to express normal behaviour and freedom from pain and distress are basic principles in this kind of monitoring.

Digital Agritechnology. https://doi.org/10.1016/B978-0-12-817634-4.00010-0

Why automated health management?

There are benefits to automated cow health monitoring to farmers, retail and consumers, as well as governments. Healthy cows are more efficient than sick animals in producing milk with minimal pollution to the environment.

The cost-benefit to automated cow health monitoring for a farmer is often hard to quantify. Early diagnosis or even prevention of disease lets the cow reach its full potential and highest production with little loss of profit. Mottram (1997) prioritised the hierarchy of costly health events as parturition, mastitis, metabolic conditions and lameness as the causes of major impacts on cow health and this chapter will follow the same order.

Illness in dairy cows is relatively rare, and hard to simulate experimentally for practical and ethical reasons, which means that researching disease monitoring is very difficult. In particular, where sensor data are measuring something invisible to humans, such as rumen pH, no data has been available until very recently and there has been a delay in teaching veterinary and animal science that uses data from sensors. Numerical analyses have to be a retrospective analysis of events prior to a subsequent clinical diagnosis, and most research studies are of short duration on limited numbers of animals. It would be unethical to deliberately create disease conditions merely to test sensors.

Rutten et al. (2013) conducted a very detailed structural analysis of scientific reports of sensor systems for dairy cows and presented effectiveness at four levels:

1. Sensor techniques
2. Data interpretation
3. Integration of information
4. Decision making

Rutten et al. (2013) found that most papers researched mastitis, fertility and locomotion, but few did metabolic conditions. They commented that most studies lack interpretation at decision-making level, which is mostly due to the lack of uniform description of diseases and treatment interventions. Thus, most systems provide data but not information to trigger an action.

One criticism of Rutten et al. (2013) is that it sees the benefit of monitoring cow health only in financial terms. The consumer is happier if the animal product is proven to be from an identifiable animal, which confers high status on the consumer. High status in human society nowadays comes from a sense of concern for the environment and animal welfare, thus being able to report the naturalness of the system (for example grazing) and the health and well-being of an animal is a motivation to collect this data. Installing electronic systems will greatly reduce paperwork, intrusive and biohazardous inspections to determine and present welfare status to the consumer via the supply chain in future.

Milk buyers have their own criteria that flow from the growing desire for provenance data and this has driven developments in monitoring. Modern consumers would like to be sure that healthy cows produce their milk and digital systems provide a potential for recording these parameters. Welfare monitoring systems are currently based on regular inspections, but soon it could become contractually essential for farms to be able to provide standardised reports at cow level.

Zoonotic diseases (Anthrax, Tuberculosis, Brucella, Red Water Fever, Weil's disease) have long been a concern of governments, as are highly contagious animal diseases such as Foot and Mouth Disease. State veterinary inspections and manual mandatory reporting currently monitor these, although, for example, in the Netherlands some has been replaced by monitoring data (Brouwer, 2015; Royal, 2021). Many of the automated systems installed on farms have the ability to detect anomalies, which may be the symptoms of infectious disease. These anomalies may be across the herd such as a drop in milk yield or only affect a few individual animals. Most farm records stay on a herd database at the farm and thus are not searchable on an epidemiological level. Data aggregation of this type has huge potential, but major issues remain in data security and ownership.

The One Health concept promoted by government initiatives aims to improve health and well-being through the prevention of risks and the mitigation of effects of crises that originate at the interface between humans, animals and their various environments.[1] This will drive the market for automated solutions in the future particularly in relation to infectious diseases. Integrated with the need to record health is the need to record treatments, particularly treatments that affect humanity. Digitizing the recording of anti-microbial treatments is important to reduce the risk antimicrobials causing resistant organisms to develop (CDC, 2019).

Increasingly, governments are looking at agriculture for ways to reduce environmental pollution. In agriculture, the principal sources of pollution are methane from enteric fermentation in the rumen and leakage of N compounds to water and as N_2O from cropland. Nutritional health is particularly important to minimise these emissions but this has not yet been a major motivation for monitoring technologies.

Traditional veterinary interventions were based on the farmer calling a vet when an animal was visibly unwell. The vet would then carry out a clinical examination, come to a diagnosis and cure or cull the animal. The impact of digital agritechnology is causing a paradigm shift in animal health management (Dumont et al., 2014). Data analysis at herd level can be used remotely to understand subclinical changes in disease status. The binary split between healthy and clinically diseased can be replaced by a continuum of numerical-based indicators triggering prophylactic and other interventions based on continuously available data.

Because of a greater emphasis on prophylactic techniques in management, fewer animals will be brought forward for clinical examination. This shifts the work of a vet to keeping animals healthy rather than treating sick animals. Veterinarians could in future remotely manage thousands of animals to prevent illness with interventions in management, such as adjustments in feed regimes, cleaning practices and bedding management rather than treating animals with surgical interventions or drugs.

---

[1] http://www.onehealthglobal.net.

Issues in automated animal monitoring

Classical veterinary animal inspection is full of subjective observations described in words and a trained veterinarian will use these to make diagnoses of diseases. These descriptive phrases and human sensing are hard to translate into a uniform, computerised description of a condition. The benefit of digital health monitoring enables a more pro-active use of limited veterinary resources by changing the definition of diseases from qualitative clinical judgements to numerical indices.

Prophylactic veterinary medicine is based on measuring an actionable parameter. The first phase is the monitoring of a differential away from the normal band, the second phase monitors the rate of change of differential as the disease either self-corrects or accelerates towards clinical level. Depending on the threshold, the rate of change would trigger an intervention or not. Monitoring continues after the intervention to determine whether the intervention is having an effect.

Classification of disease

Analysing large amounts of data from a herd and comparing individual animals against the mean measurements can overcome the uncertainties in classification of events and treatments. When we get a deviation from normal, we need to decide on a diagnosis. Healthy animals are equipped with a strong immune system, which is able to deal with illness continually. We need to be wary of constantly treating animals with what are in effect sub-clinical symptoms of a disease, which they can overcome easily. For this reason, the objectives of the monitoring should be to indicate animals that have symptoms for an intervention that contributes positively to animal health and welfare, whilst respecting the physiological potential of the animal.

Take as an example the simple measure of temperature. In addition to the unit of measurement (Fahrenheit, Kelvin or Celsius), the temperature data needs clarification as to where on the animal (ears, subcutaneous, rectum and rumen) the measurement is located, over what time period the temperature was measured and what the ambient conditions were at the same time. Animal activity, gestational status and breed also influence what we classify as a deviation of normal temperature. Until recently, these kinds of data have been wholly lacking for different species and breeds and their adaptations to environmental conditions. We would not expect the surface body temperature of a white Holstein cow in the tropics to be the same as a Black Highland cow in the sub-arctic Scottish highlands. Measuring core body temperature is effectively impossible in live animals, so we only have proxy values. For example, in measuring rumen temperature, the sensor needs to be averaged out to obtain a true baseline temperature linked to core body temperature of the vital organs, irrespective of activity and ambient temperature. Measuring rumen temperature is highly affected by activity; it shows sudden rapid drops in temperature when the animal drinks cold water. The unexpected benefit is that the sensor cannot only measure temperature, but also provides useful information about drinking behaviour. Most temperature measurements are being studied to manage the conditions at housing, for example, the ETIC model (Wang et al., 2018). The physiological response of the animal may be actually better and easier to measure method. Breathing rate is relatively easy to determine from bioacoustics or visual indicators, although again whether the animal has been exerting herself needs to be factored in.

The dynamics of the change of health status also need to be considered. Some events such as dystokia at parturition or coliform mastitis can develop from no symptoms to death within a few hours, whilst others such as change in body condition occur over months.

Sub-clinical diagnoses

Post et al. (2021) defined an important difference between low-risk events that occur sporadically on single days (mastitis) in comparison to chronic conditions that persist for days (nutritional disorders, lameness). This difference can also be used to break down the diseases into different causative agents so that mastitis, for example, would be redefined into each pathogen pathway. Even for relatively common conditions of disease in dairy cows (mastitis and lameness), the incidence of a clinical case is unpredictable and we have limited information about the parameters to measure prior to a clinical case. There is now a mass of data coming from sensors and very little is ever analysed.

A feedback loop is needed which would be triggered when a clinical case is identified and treated by a human and then all the datasets from sensors in the days prior to the case could be analysed and used to build a generalised predictive model. This form of machine learning is a very powerful tool to continuously improve algorithms; the key is standardising the formats of data, curating human input to standard terms and not overloading the veterinary technician or herdsman with more than can be dealt with daily.

Sub-clinical infectious disease is mostly dealt with by the healthy animal's immune system and human intervention, particularly with anti-microbial pharmaceuticals, and is to be avoided where possible. The key to using sub-clinical disease markers in the integrated data is identify the trends that indicate that the disease is on track to self-heal. A recent example of this approach in mastitis detection is by Hogeveen et al. (2021) breaking down disease definition into actions — immediate treatment, those not needing immediate treatment and those to be treated at the end of lactation.

Cow identity: the essential first component

In the 1970s, the author worked on a traditional Scottish beef farm. 30 black Galloway crosses out-wintered on rough pasture with little supplementation. To the untrained eye, they all looked the same, but the herdsman had names for all of them and kept a veritable family tree of each cow in his head. They did have pedigree marks tattooed into their ears, but since these animals were effectively wild, even when restrained, reading the ears involved considerable person risk. These practices have had to change as herds have become larger. The needs of biosecurity and stock control have brought in unique identity systems with national level coverage in many territories.

Security of identity is of vital importance in cow management. In the EU and many other territories, every bovine has a mandatory unique alphanumeric identity registration with a government agency or pedigree board. The code structures generally conform to ISO 11784:1996 Radio Frequency Identification of Animals, which describes a 64 bit code structure expressing 15 decimal digits. The first 3 digits are used as a country or manufacturer code. The remaining 12 digit number is a unique animal identifier given at birth and can be used to trace the location of the animal for disease control purposes throughout its life. Records have to be kept of animal locations and movement off the holding. Automated monitoring of dairy cows is predicated on individual animal identification and any identification system must link the recognisable ID of an animal on a farm (typically 0–9999 or a name) to the national database number. Until the potential for remote recognition using cameras is realised (Kumar et al., 2017), the only viable source of identity for automated monitoring is electronic identification (EID).

The location of permanent ID systems and sensors on dairy cows is limited to ear tags, implants, collars and boluses. Ear tags are used for mandatory non-EID and EID. One company Smartbow (now part of Zoetis) have demonstrated location and movement sensing with powered ear tags and this system design may become more common as electronics develops lower power versions. However, the future of ear tags in developed countries is possibly in doubt given the desire to reduce 'mutilations' for animal welfare purposes. Apart from the obvious wound from attaching a tag to the ear, tags are frequently ripped from ears causing an infection risk and possible loss of function (fly swatting and hearing).

EID devices are low cost and widely used in other industries and can be mounted on the animal in the form of ear tags, collars, leg tags and boluses. The smallest tags use passive electrical generation to generate an ID signal when energised by an inductive antenna. This inherently limits their range of operation to less than a meter from an energised antenna, for example, when in a milking parlour or automated milking system. Newer Ultra High Frequency tags are powered by battery, can store data and send radio signals at longer ranges of potentially several kilometres. The battery life may be the limiting feature of these systems although claims of over 5-year operation have been made. These small tags are light enough to be fitted to the ears.

Collars are a traditional way both to catch and restrain a cow and also to attach bells which provide an audible location signal still in use today. Collars permit heavier monitoring devices (up to 0.5% of body weight) but may need adjustment for the seasonal changes of weight that affect the neck diameter. Most farms find that the wear and tear of fabric collars limit their lifetime, but the electronic modules can be moved from collar to collar and can be used on multiple animals to make the most of the capabilities, for example, for detecting oestrus. Good digital records must be kept if collars are moved from cow to cow and most EID systems have multiple programmable fields so that the device ID can be linked to the animal national ID and farm ID by the database. Even so, numerous problems occur with cow identity due to the multiple numbers, tags being lost, transcribing errors in hand written systems, fraud and false tagging.

Insecurity of identity is not a problem for boluses, which are retained in the rumen reticulum for the lifetime of the animal, if the specific gravity is greater than approximately 1.5. However, the animal tissue inhibits radio signals and inductive antennae have not proved popular for bovine ID purposes using boluses. LoRa radio does give greater radio range but this is unlikely to exceed 200 m in the author's experience of rumen telemetry.

Types of sensor systems
Wearables

It was a natural progression to add sensors to the visible ID systems mounted in tags, collars, leg straps and boluses. The simplest of these were movement detectors based on tilt switches that could count steps. Since the advent of low cost tri-axial accelerometers after 2000, tilt switches have been replaced by a combination of accelerometers and models to convert the change of orientation of the sensor into counts of steps, which can then be converted into behavioural measures such as oestrus detection, lying and feeding behaviour. These behavioural statuses can then be converted into a diagnostic for various conditions (Figs. 6.1 and 6.2).

Mottram (2011) patented the measurement of lying and standing behaviour using a tri-axial collar based on measurement of the angle of the collar to the vertical at periods when the vertical amplitudes are low. As cow does not lie with her neck vertical the z-axis, which normally measures turning behaviour contains a gravitational component so that the collar can even measure whether the cow is lying on her left or right side.

Collar-mounted electronics provide a good place to mount radio systems, giving elevation and lower radio attenuation from the body mass of the animal and its neighbours. Drysdale et al. (2008) modelled the radio fields for collar-mounted sensing systems, and concluded that the occlusion of signals by the bony poll of

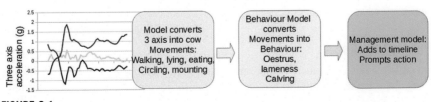

FIGURE 6.1

Models are needed to convert accelerations into movement data and then into behavioural data on a timeline to compare to descriptions of indicators of illness.

FIGURE 6.2

The three axes of acceleration can be resolved by vector trigonometry to determine the orientation and movement of the neck.

the cow meant that static antennas should be placed about 4 m above the cow pen in a location where they were preferably behind the animal.

The engineering issues relating to wearable sensors relate to power management and the trade-off between embedded models and radio transmission. Data communication bandwidth and power consumption need to be optimised. There are also trade-offs between bandwidth, frequency and radio range. More data can be transmitted faster at Wi-Fi frequencies in Gigahertz but range is rarely more than 30 m, which may be acceptable for housed animals, but is difficult to maintain while on pasture. Power allocation between radio transmission and computing is a major factor in device design.

The models for diagnostics work best when integrated with different data sets, which is best done in a cloud or centralised location. This is a classic trade-off between edge computing by small computers with limited data and cloud computing with massive resources. Models can be mechanistic, using vector Pythagorean methods to resolve directional forces, tilt angle, etc. Modern pattern recognition techniques can then be used to map against visible behaviour and a retrospective analysis of a diagnosis. The disadvantage of these neural network and artificial intelligence techniques is that explaining why they work is detached from the scientific reductionist method. In addition, it is hard to create a clean universal training set of data. Presenting the similar data to farmers and stakeholders becomes therefore somewhat of a challenge.

Image capture

As cameras have become cheaper and are already fitted as standard in robotic systems, there is the potential to use them to monitor identification, behaviours, rumen activity, breathing rate, location and body condition score (BCS). Cameras have already been developed for monitoring body condition (Mottram et al., 1997, Hansen et al., 2018). Song et al. (2019, p. P9076) described a system for monitoring the flanks of the cow using dynamic image capture to determine rumen motility. These show great potential for the future as quality of cameras and the capability of software develops. Data extraction requires a great deal of processing but this can be achieved by cloud-based systems.

Thermal imaging

The silicon charge couple arrays that are the basis for digital cameras can be tuned to detect the infra-red spectrum and this forms the basis for systems of thermal imaging that detect variability of surface temperature. There are some technical issues about noisy signals due to the difference in thermal performance of different pigmented skin and reflected sunlight, but with suitable location and software filters, these can be overcome at least in research environments.

Frondelius et al. (2021) concluded that thermal imaging has a good application in two areas. It can be used post-calving to detect hypocalcaemia (Milk Fever) where the cow's body temperature drops rapidly. Thermal imaging can also be used to pick up hotspots of infection, for example, in cows' feet suffering some forms of lameness or in the udder for mastitis.

Biosensors

Markers of disease or biomarkers can be monitored and measured in emanations from the animal such as breath (Turner et al., 2012), saliva, sweat, odour and milk (reviewed by Neethirajan, 2017). Only systems for milk analysis of hormones and acute phase proteins (APPs) have yet been successfully marketed. As milk is a dairy farm's end-product, it also offers a direct link to provenance for the milk processors. Many bodily emanations have variable concentrations of the chemical being detected and thus this concept is best used where the signal is effectively dichotomous, such as antibodies to a given pathogen either being present or absent. The detection of biomarkers for pathogenic disease will be highly dependent on each disease and each will have a different potential, in both engineering success and economic contribution to a farm.

Biosensor is a generic term for a wide range of sensing technologies and has been widely applied to different compounds secreted by animals. Most practically, they have been applied to milk. The basic principal of all biosensors is that the compound of interest causes a reaction in the sensor, which may change its colour or electrical properties. A process of transduction measures the signal of the response of the sensor, which is converted to a digital signal read directly by a computer. Subsequent modelling could take place on the edge, in or near the sensor, or the signal could be combined with other data in the cloud in more complex modelling. The potential to use them for individual animal analysis is immense and one can imagine that when assays are readily and cheaply available for DNA markers of antigens or antibodies against specific diseases found in milk, then routine testing for a number of infectious diseases becomes possible.

Simple compounds such as urea and BHB can already be measured by enzymatic reactions which induce a colour change (Mello et al., 2020) or by an electrochemical signal, like the widely used blood sugar sensors used by human diabetics (Wikipedia, 2021).

Where a biochemical biomarker is known, such as a hormone or specific protein markers, then lateral flow immunoassay (LFA) is probably the most appropriate technology to identify them (Koczula & Gallotta, 2016). Colorimetric LFAs have become very well known during the Coronavirus pandemic to detect antigens. The principal is to wash the sample swab with a buffer and extract a liquid, which is applied to end of the sensor. A colour change captured by eye or a camera in a light controlled box will identify the biomarker concentration. Lateral flow immunosensors require small samples for analysis, typically less than 200 μL. The sensors are delicate one-shot devices that need extensive protection in packaging, to maintain optimal temperature and humidity, which may be difficult in a milking system on farm. They can only be used once and have a shelf life of merely months. When these problems are overcome, for example, in the DeLaval Herd Navigator, they can, for example, be used routinely for progesterone analysis for fertility status monitoring.

Monitoring cow health

Healthy production includes all stages of the life of a cow. Every lactation is the start of a new dairy cow as a calf, as well as the propagation of an already producing dairy cow. Cows can live far longer (nearer 20 years[2]) than current average age on farm of 5 years (de Vries & Marcondes, 2020). A cow's whole lifecycle is intertwined. Thus, the health prospects of a herd cascade through generations over decades from simple prophylactic interventions made possible by good monitoring technologies and management interventions.

It all starts with the health of the pregnant animal. Epigenetic studies (Singh et al., 2012) suggest that the good health and productive potential for the future offspring of the unborn foetus will be predicated on the health status of the pregnant animal at conception. As an embryo, the female calf develops the oocytes that are the potential new embryos (Britt, 2008). Thus, health management of the pregnant mother will form the oocytes in the foetus for the future calf's lifetime supply of potential embryos.

The health of the new-born calf is highly dependent on an easy parturition and immediate good quality colostrum. This is most easily achieved by ensuring pregnant cows are in the correct condition for calving. So, for example, a BCS system could ensure late lactation and dry cows are not too fat and this reduces risk of calving problems and leads to healthy calves. Measuring colostrum and providing the best quality to the calf will ensure the best start in life (Fig. 6.3).

---

[2] Smurf, a Holstein cow in Canada, passed 12 lactations with a total yield over 216 tonnes. https://www.guinnessworldrecords.com/world-records/greatest-milk-yield-by-a-cow-%E2%80%93-lifetime.

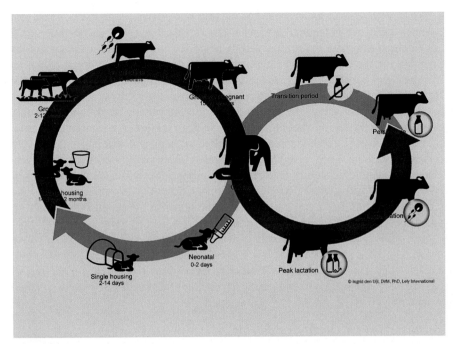

FIGURE 6.3

Lifecycle of a cow.

Monitoring technologies for calves will have to rely on remote sensing for growth (Song et al., 2018), condition and fertility, while monitoring of adult cows will in addition be focused on production, which can be monitored through milk. Monitoring body morphology and function, as well as fertility and general health, is still very important in adult cows.

Parturition (calving) monitoring

The transition from dry to lactating through calving is the most dangerous time of the cow's annual cycle needing careful management. A survey study of 10,000 cows in Finland by Sarjokari et al. (2018) showed that 5% of cows died because of calving problems. Similar numbers have been consistently reported for decades across all territories, although there are variations in genetics and management systems. Management of the cow before and after parturition (transition cow management) is critical to the subsequent health and fertility of the cow and calf.

Better prediction of the time of calving would enable prophylactic measures to adjust diets and management to reduce mineral nutrition problems. Although the mean length of gestation can be added to the insemination date to predict the date of parturition, the sex and breed of the foetus and other variables mean that there is usually a wide distribution around the predicted date. For optimal management, a farmer will need to know when the calving has started, if there are issues and whether assistance is required, and when the calf has been born successfully.

The foetus initiates parturition, and there are few physiological indicators of the pregnancy in the dry state that can be measured non-invasively, leading to focus on monitoring of behavioural data as the cow herself senses the onset of calving. Since the majority of calvings proceed normally and human intervention is both expensive and potentially damaging, any system for monitoring calving needs to identify only those animals that are not able to complete the calving process without assistance.

Gundelach et al. (2008) monitored the effect of human monitoring and intervention in calving on a 463 cow German commercial farm and concluded that there were few negative impacts of improved monitoring. Interventions should be considered when second stage of labour lasts longer than 2 h. Insufficient monitoring around parturition had a negative effect on the duration of second stage of labour, and thereby, indirectly on perinatal mortality.

In addition, monitoring the calf is just as important. 8.5% of perinatal calves perish, which can largely be attributed to issues around calving (Santman-Berends et al., 2019). The birth of a dairy heifer is a very important event for the rest of her life; calves that are born with difficulty will not fulfil their full genetic potential as lactating cows.

Technologies for monitoring calving

Calving detection is an area where new technology is being developed to show a major benefit on farms. Behaviour of a cow is indicative of the start of calving and whether there are issues that need assistance. This behaviour can be monitored by various tracking devices like cameras or accelerometers. Miedema (2009) showed that changes in a 'restlessness' index counting the number of times a cow changed from lying to standing and the reverse in the final 6 h before calving (appearance of contractions) can be used to detect the onset of calving when compared with the individual cow's 'restlessness' index during the dry period 21 days before calving. This work also attempted to predict problems with calving (Mediema et al., 2011) but found no correlation between behavioural change before calving and dystokia once it had started.

Maltz et al. (2011) showed that measurement of lying time, number of steps, number of lying bouts and movement through the barn could be modelled to predict calving time up to 48 h before calving with 90% true positives although there were 15.6% false alarms.

This behaviour was also seen in a study by Adrie Meeuwisen (as yet unpublished, Lely Industries NV) based on tagged camera data. In this study, it was possible to determine dystokia in 11 out of 23 cows based on different frequency and duration of behaviour. During a smooth calving, a period of restlessness was followed closer to calving by the tail rising and contractions. Very often, around the time of driving out the calf, one or more legs and the head would be stretched for a short time and contractions were at a maximum (Fig. 6.4A). A cow going through a difficult calving would show higher frequencies of these behaviours in a more erratic way (Fig. 6.4B).

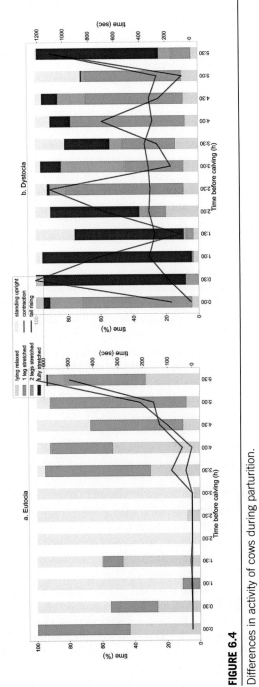

FIGURE 6.4

Differences in activity of cows during parturition.

Fig. 6.4 shows the behaviour of a cow with a smooth birth, whilst Fig. 6.4B shows the behaviour of a cow that needed assistance with calving.

Rutten et al. (2014) found that a model based on measures of rumination, activity and temperature on 417 calvings was predictive but with high false positive and negatives. Borchers et al. (2017) subjected similar behavioural data from existing cow-mounted technology with two monitoring devices (HR label and IceQube) on 53 primi and multiparous cows to machine learning to attempt to predict calving time. By combining data about lying, steps, and rumination in a neural network analysing over 2 hour periods in the 8 hours before calving, they were able (on the same data set) to show a capability of predicting the start and speed of calving 82.8% specificity and 80.4% sensitivity. The same data set was analysed later with different machine learning techniques by Keceli et al. (2020) who claimed to have improved toward 99% specificity and sensitivity. Unfortunately, these algorithms were not validated on groups that were not part of the development set. If more categorised data like that of Borchers et al. (2017) were available from an open repository, more progress could be made in building reliable models. Other proxy methods of detecting the onset of calving also largely rely on behavioural indicators such as the Smartbow (Zoetis) ear tag, which is permanently attached to the cow's ear. Krieger et al. (2019) demonstrated an algorithm to use this device to predict calving but only achieved 74% sensitivity 1 hour before the onset of visible signs.

More invasive methods have involved inserting temperature probes into the vagina (Burfeind et al., 2011) and detecting a drop in temperature in the hours preceding calving, and the final drop would occur when the temperature probe was expelled with the foetus. Ouellet et al. (2015, p. P1539) evaluated devices measuring behaviour and vaginal temperature as a predictor of calving both singly and in combination. The best predictor of calving within the next 24 h was a drop in vaginal temperature $\geq 0.1°C$ at sensitivity and specificity both of 74%, whilst combining these data with behavioural measures increased both the sensitivity and specificity to 77%. Insertion of devices into the vagina may raise risk of infection and still needs a timely intervention and may only be appropriate for high value animals at risk of dystokia.

One commercially available device is Moocall (2021), which is strapped to the tail head and emits a warning signal via the SMS system when the tail becomes rigidly extended as a sign that calving is taking place. It has the disadvantage that it requires attachment some hours in advance of parturition, which implies a degree of observation and management, which is what monitoring is meant to reduce. In one experiment, only 14% of the attached sensors stayed in place for the duration of a test and had to be re-attached if the tail was swollen or if sore, removed. Because of variability of the interval at which it was attached before calving, sensitivity varied from 19% to 75% and specificity from 63% to 96% (Voß et al., 2021).

As shown by Gundelach et al. (2008), the extended duration of the second stage of parturition is most likely to cause problems and calf mortality. Therefore, calving monitoring should focus not just on prediction of calf arrival but also on the calving process itself. If a calving is proceeding normally, the womb and uterus begins a series of contractions to expel the foetus into the environment. As parturition proceeds, the contractions will get stronger and more frequent until the calf is expelled. If the contractions become weaker and further apart, then intervention might be necessary, because the cow may be suffering from dystokia and/or hypocalcaemia.

During calving, the contractions can be detected by electrohysterogram measurement with a belt strapped around the cows hind quarters on which were mounted adapted ECG probes (unpublished, Author's collection). However, this approach, like Moocall, requires management intervention and cow handling and has not proceeded commercially.

In future, we can predict that developments in image processing to detect contractions and behavioural change at parturition will become possible. This is likely to be the most beneficial use of research resources as it provides the potential for a completely automated non-invasive system (Fig. 6.5).

Mastitis

Mastitis is an inflammatory response of the mammary gland to a challenge, usually bacterial. Mastitis can occur at any time in lactation but is most common when animals are under stress and also occurs during the dry period. Mastitis detection has to be completed during milking to enable contaminated milk to be separated before it enters the food supply chain. Early detection also means that the disease can be treated in a timely manner. The challenge in mastitis detection is that each pathogen causes a different inflammatory response. During conventional milking, a traditional role for humans is to detect clots, behavioural changes, discoloration and swollen or heated udders. Robotic milking requires sensor technology to achieve the same results. Robotic milking is standardised by ISO 20966 (ISO, 2007), but the requirement to detect abnormal milk is only described in the Annex C (Informative). Currently, no definitive method covers all inflammatory responses. Reinemann

FIGURE 6.5

An electro hysterogram measuring system developed by the author.

and Helgren (2004) concluded that robotics provide sufficient information for motivated dairy producers' to achieve national milk quality standards. Helgren and Reinemann (2006) reviewed the methods commonly in use in automated milking, which are largely based on colour change and conductivity, with associated algorithms.

Mastitis detection is still developing, but the complexity of the disease and the need to develop non-antibiotic treatments mean that innovation is continuing. The main approach is to use conductivity, behavioural change to detect developing sub-clinical conditions and then to confirm diagnosis with an somatic cell count (SCC) test during milking. Frondelius et al. (2021) exploring bovine applications of thermal imaging concluded that it was not reliable for mastitis detection.

Conductivity

Conductivity has been used to detect mastitis since the 1970s, but it lacks sensitivity and specificity to all inflammatory responses. Milk conductivity is attractive as an engineering system since it needs no reagents and can be cleaned in place. To improve on single sensors, an array of conductive sensors with different cross-sensitivities was used by Mottram et al. (2007). A total of 67 samples of milk from both mastitic and healthy glands were measured. It demonstrated that the multi-sensor system could distinguish between control and clinically mastitic milk samples. The sensitivity and specificity of the sensor system (93% and 96%) showed an improvement over single electrode conductivity (56% and 82%). The multi-sensor system offers a novel method of improving mastitis detection; however, manufacturing issues to improve sensor consistency need to be addressed.

Sloth et al. (2003) demonstrated that a multivariate approach (milk yield, protein percentage, fat percentage, lactose percentage, citrate percentage, SCC and two electrical conductivity parameters included 821 cow level observations) can detect udder conditions and aid udder health. As endotoxin bacterial infections progress very rapidly towards severe illness and even death, systems are necessary to detect sub-clinical signs. Behavioural change and thermal imaging of udders (Hovinen et al., 2006) have been demonstrated to have potential particularly in robotic situations.

Somatic cell counting

Milk quality standards for mastitis are expressed in SCC which is measured in laboratories from a bulk milk sample. The California milk test (CMT) has been used successfully as a manual cowside test since the 1970s and has been automated as CellSenseTM and offered as an option on milking systems by LIC Automation (2015) and Lely. CMT uses a reagent which disrupts the cell membrane of somatic cells present in the milk sample; the DNA in those cells reacts with the test reagent by changing the viscosity of the test solution. Neitzel et al. (2014) tested sensors for CMT and showed that the significance between sensor variations could be improved by a different calibration method. Rossi et al. (2018) used a somatic cell system which gave accurate results for *S. agalactiae* with very high correlations against bacteriological tests.

A different method is used in the DeLaval online cell counter (DeLaval, Tumba, Stockholm County, Sweden, 2015) where a reagent is mixed with a sub-sample of milk to enable an image capture system to count cell nuclei. The cost of reagents militates against the cell counting tests being used at every milking but can be deployed when other indications (conductivity, discolouration, temperature rise, behavioural changes) require validation.

Acute phase proteins

Eckersall and Bell (2010) reviewed APPs in veterinary medicine as biomarkers of inflammation, infection and trauma. In cattle, haptoglobin and $\alpha1$ acid glycoprotein and haptoglobin and serum amyloid A have proved valuable biomarkers of disease, respectively. In dairy cattle, haptoglobin and a mammary-associated serum amyloid A3 isoform, produced by the inflamed mammary gland during episodes of mastitis, have great potential as biomarkers of this economically important disease. Biosensor technologies to detect APP lactate dehydrogenase are fitted as standard on the DeLaval (2021) VMS system and can be used in combination with other data to give an accurate measurement of a persistent mastitis (Fogsgaard, 2015). The disadvantage of biosensors is that they are one shot disposable systems, so they are probably too expensive for use at every milking to detect what are relatively rare events although the huge increase in production capability of lateral flow immunosensors during the COVID-19 pandemic of 2020/2021 will probably reduce the price significantly in future.

Metabolic disorders

Nutritional management is key to healthy productive animals, and traditionally this has been largely an open loop control system, which is only adjusted when production or health fails, sometimes with considerable delay after the feeding mistake. Traditionally, monitoring nutrition has been achieved by milk yield and composition. Body weight and BCS have been used to identify the success of a feeding program, but these systems have a long feedback loop. Homeostasis causes the cow to adjust her output to the ration offered, milk yield may decline, but the markers of energy and protein metabolism (BHB and urea) may stay within healthy bounds.

The main inputs of feed to the cow are energy and protein, but these have to be managed in parallel with fibre, micronutrients and minerals, which come from many different sources. In reality, we are feeding not the cow, but the rumen microbiome which is more complex. Cows are ruminants with a four-chambered stomach (or four stomachs) with a volume up to 120 L. The main ventral sac of the rumen is a fermentation chamber where the rumen flora break down fibrous cellulose from plants into compounds for digestion in the other rumen chambers (masum and abomasum) and the hindgut to extract nutrients from the feed stream. The rumen flora is very diverse with specialised organisms and enzymes competing and cooperating.

What we know about the rumen comes from invasive studies using fistulated cows from which samples can be taken and new compounds tried out to identify the breakdown process. Because of the expensive invasive surgery needed, the careful management of a vulnerable animal and the disruption of the true anaerobic nature of the rumen, fistulated animals will almost certainly be replaced by telemetry and modelling in future.

Grazing cows are crepuscular feeders and this is overlaid by the time at which feed is offered to housed cows which stimulates more feeding behaviour. A common condition of high yielding dairy cows is acidosis, which was traditionally defined qualitatively as extended periods of low pH. Cows love eating and they can easily overeat (for example, if they escape into a feed shed) and drive the pH so low that the digestive bacteria die and the rumen becomes congested: death then follows quickly. More commonly, cows can eat excessive amounts of concentrates by either selecting from inadequately mixed feed (sorting) or over-feeding at milking time; both have a major negative effect on rumen acidity.

There are three techniques to determine metabolic status: rumen telemetry, which measures the status of the digestive process in the rumen directly, behavourial and rumen movement analysis with externally mounted telemetry or cameras and milk constituent analysis, which can give an indication of imbalances after the cow has processed the feed.

Rumen telemetry

The fibre-digesting rumen organisms thrive at higher pHs (above 6.3 pH) and the starch digesting bugs are most active at lower pHs down to about 5 pH. When the diet is heavy with starch and low on fibre, digestive organisms can push the pH even lower. Rumen pH is known to be highly variable in time, with up to 2.5 pH range through the day and varying spatially up to 0.5 pH units from top to bottom within the rumen at any given time (Gasteiner et al., 2010).

Measuring rumen pH by telemetry is a direct way of ensuring that there is a balance between the behavioural routines of the cow and the continuous nutritional requirements of the rumen flora (Mottram, 1997).

Historical veterinary methods for measuring ruminal pH in commercial cows were based on either rumenocentesis or through use of an oral sampling tube (Tajik & Nazifi, 2011). Both methods are invasive and can only gain one data point from an imprecise location within the rumen. Small studies appear to show that estimations of acidosis incidence based on rumenocentesis (Atkinson, 2013) are too high, whilst Jonsson et al. (2019) showed that single point testing has limited diagnostic utility in comparison. Direct measurement of the rumen pH by wireless telemetry was reported by Mottram et al. (2008) and Gasteiner et al. (2009). The wireless telemetry bolus allowed continuous recording of data from a fixed location within the rumen-reticulum, thereby overcoming the variability in data.

A definition of acidosis or more accurately Sub-Acute Rumen Acidosis (SARA) based on rumen pH dynamics was developed by INRA (Villot et al., 2018) who also showed the need to process out noise caused by sensor drift. The wealth of data these boluses provide has yet to be integrated into mainstream nutritional thinking which was based on theory before direct continuous measurement of intact animals was available.

Smaxtec, Well Cow and eCow[3] boluses measure pH and temperature and store the data for download by radio. Some boluses download data when the cow is within range of fixed monitoring stations; others used a handheld device to enable a human to download whilst inspecting the cow (Fig. 6.6).

The chief limitations of the pH technology are that boluses are too expensive to be used on every cow and have short lives due to pH sensor poisoning. Additionally, once the bolus containing the sensors is deployed, it cannot be retrieved for servicing and analysis. Boluses may have either ion-selective field-effect transistor pH sensors or glass electrodes. Ion selective field-effect transistors appear to degrade faster than glass electrodes, which are larger. Lifetime is usually limited by the stability of the reference electrode.

Rumen boluses have also been sold to measure temperature only, these are considerably cheaper than pH sensors but do suffer from a lack of models as to how data can be used. Rose-Dye et al. (2011) showed that temperature boluses detected infection challenge in steers (Fig. 6.7).

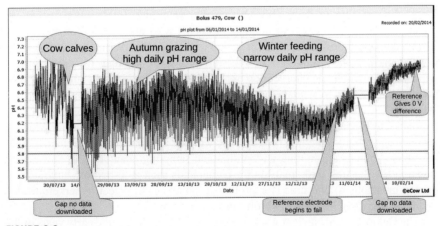

FIGURE 6.6

A bolus provides a wealth of data over its 6-month life. This bolus was inserted before calving and the abrupt change in feed at calving can be seen. Then there is a high daily range of pH as the cow was at grass and had twice daily feeds of concentrates as well. When winter TMR feeding started in October, the daily range of pH reduced. Later, an artefact caused by the reference electrode becomes poisoned and the voltage differential reduces to zero which is reported as 7 pH.

---

[3] Both Well Cow and eCow companies were founded by the author (Mottram).

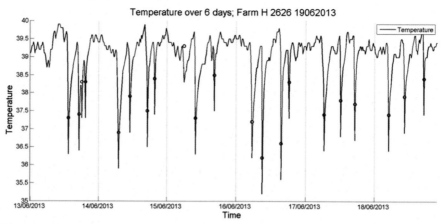

FIGURE 6.7

A temperature profile measured in the rumen reticulum of a dairy cow. The data points are the 1 min values averaged every 15 min. The steep drops indicate drinking events and can be used to determine whether sufficient water is available for the animal.

In-line milk sensing of nutritional parameters

Metabolic profiling of dairy cows began as a veterinary intervention before the 1970s (Adams et al., 1978) to identify problems with nutritional management. It involved taking blood from a sample group of cows and performing a detailed laboratory analysis and a report back to the farmer a few days later. It was sometimes deployed as a diagnostic intervention or as a routine tool to improve nutrition. Blood analysis can be used to validate the results of measuring milk parameters of nutrition status.

Laboratory analysis of milk chemistry has a long history (Schultz & Megess, 1959). Regular milk sampling became easy with the introduction of automated milking where animals are routinely identified and sampling is automated. Subsequent research tried to link milk analysis for compounds such as β-hydroxybutyrate (for energy metabolism) and urea (for protein metabolism) to detect hyperketonaemia a clinical nutritional condition (Mottram et al., 1999).

The DeLaval Herd Navigator was progressively launched in the early years of the 21st century and monitors five inline milk constituents with 'dry-stick' immunoassay for progesterone and colorimetric enzyme sensors for β-hydroxybutyrate, urea and lactase dehydrogenase (Mazeris, 2010). With the addition of behavioural information and other data, this gives a comprehensive analysis of fertility, protein, energy status and mastitis. The disadvantages are the expense and size of, what is in effect, a temperature and humidity-controlled laboratory. Typically, it is attached to a robotic milking system. The benefits were calculated to be between €250 and 350 per cow per year, most of this comes from the fertility management using progesterone (Leonardi, 2013) who also stated that the ketosis and mastitis detection features were not beneficial on their test farm. An indication that fertility is the most important element is that in 2018 DeLaval launched the rePro system, which monitors only progesterone using a lateral flow assay deployed in a cassette from which a covering film is peeled back for each sensor. From the author's experimental program in the 1990s, inline milk analysis for nutritional management yields limited benefits for high cost.

Weight, body condition and body morphology

As lactation progresses, BCS changes along with the weight of the cow first falling then rising as lactation progresses. Auto-weighing is offered by a number of manufacturers as a standard accessory for milking parlour exits and walk through passages linked to EID. A different approach is to instrument the floor of the robot to detect weight. Alawneh et al. (2011) showed that daily variability (SD 17 kg) of weight could be overcome by taking an 8-day rolling mean. The standard deviation of daily live weight (LW) measurements across parities was 17 kg, on average. A near perfect association between LW measured statically and walk over (concordance correlation coefficient 0.99, 95% CI 0.99 to 1.0) was observed and recommended that LW are recorded on a daily basis to allow changes in physiological status such as the onset of acute illness or oestrus to be detected in a 7-day management review.

An alternative approach is to measure BCS using image processing. This has the potential benefit that cameras are inherently cheaper and simpler to maintain than mechanical weigh cells. Bewley et al. (2008) were able to show a good correlation between BCS and digitally processed static images. Anglart (2014) was able to show a good correlation between automatically captured digital images and weight on Swedish Red cows but had difficulties with black and white pigmented cows. An alternative approach by Halachmi et al. (2013) was immune from pigmentation by using thermal imaging showing a Pearson correlation of 0.94 with manual BCS. The DeLaval VMS system is now available with an image based-BCS system.

Image processing is capable of more than just emulating human sensing. BCS is known to be unreliable between observers (Morin, 2017), so any technique using BCS as golden standard will suffer from that. Cameras are able to determine volume and morphologic changes on multiple areas and compare them to historic images. These developments may lead to a better understanding of the relation between body changes and adverse health events.

Mobility and lameness

Lameness has come to the fore as a major problem in the dairy sector and a means of informing the herd management and supply chain as to issues in any given herd has become a priority. The real requirement for the industry is not just to detect lameness but also detect sub-clinical changes in mobility score so that interventions can be made earlier. The requirements of the milk buyers are to be able to show objective numerical data across a whole herd that shows that lameness is being reduced.

Mobility score was developed as a manual tool to aid in classifying herd lameness and identifying trends. However, manual systems are inherently subjective and widely criticised. A number of approaches have been tried to automate mobility scoring (walk over load cells, leg load during milking, back posture with three-dimensional (3D) cameras, dynamic analysis, behavioural analysis and time of passage). Alsaaod et al. (2019) reviewed various automated methods of lameness detection as being kinematic, kinetic and indirect methods and the majority were evaluated by comparison with visual locomotion scoring systems. A major problem in this area is the relatively low number of lame cows (mobility score 3 or more) and the need for statistical strength given the larger number of low mobility score cows. Although there have been many scientific studies to develop sensor techniques and algorithms to detect lameness and/or lesions, there have been no longitudinal studies to study the effect on the amount of lameness in a herd, which is multi-factorial problem.

There are a number of options for measuring lameness and even mobility score and this is an area for exciting future developments.

Walkover weigh cells

Tasch et al. (2001) patented a system where a cow walks over grid of load cells that measure forces, and algorithms that map these data to produce a lameness score automatically. The system has been available commercially as Stepmetrix and was tested by Bicalho et al. (2007) to evaluate visual locomotion scoring and Stepmetrix locomotion scoring for detecting painful digit lesions. When performed by trained veterinarians, visual locomotion scoring performed better than Stepmetrix in detecting cows with painful lesions (Ackerley, 2014).

Individual foot load analysis

Pastell et al. (2009) used load cells in a robotic milk system with 37 cow scores to measure leg load during milking and classified 96.2% correctly as mobility score (0,1,2) against 7 lame (mobility score 3 or above). The technique was developed into a real-time monitoring system using wavelet analysis and achieved a correct classification of 96.2% of lame cows (Pastell et al., 2010).

Kinetics — tri-axial accelerometers

The cow collar and leg tag with tri-axial accelerometers provide an available device that has been adapted by many researchers and companies to provide a dynamic analysis of mobility score (Alsaaod et al., 2019).

Van Hertem et al. (2013) used existing data sources from milking and activity monitors to discriminate between 44 lame and 74 sound animals. A logistic regression model was developed based on the seven highest correlated model input variables. After a 10-fold cross-validation, the model obtained a sensitivity of 0.89 and a specificity of 0.85, with a correct classification rate of 0.86 demonstrated that existing farm data can detect clinically lame animals. A much larger study by Kamphuis et al. (2013) on 4904 milkings of grazing cows with 292 lame cows measuring weight, milking parameters and pedometers concluded that detection performance was not sufficiently accurate for commercial use.

Kinematics — imaging systems

When a cow's mobility score increases, the cows back arches and step length decreases. Two-dimensional imaging needs an unobscured view of the side of a single cow and this is hard to achieve in practice on farms. Viazzi et al. (2014) compared two imaging methods to measure the degree of arching with 2D and with a 3D approach using the low cost Kinect camera developed for games playing. The accuracy of the two systems in detecting arched backs was comparable at 90%.

Time of passage measurement

Mottram and Bell (2010) showed that capturing the time taken for an individual cow to pass along a passage is closely correlated to the 5-point mobility score. As long as the cows are allowed to walk at their own speed, cows with a higher mobility score (most lame) took longer to proceed through a raceway. A follow-up study by Martinez-Ortiz et al. (2013) showed that the speed of cows could be measured by remote video tracking by segregating cows even in wide races.

Integrated monitoring

Sensors are only components of a complete system. Automated farm assessments need much more development and test before they can be trusted to replace the human judgements currently conducted. Frost et al. (1997) made a key point in the development of monitoring systems; animal health is a multifactorial system that rarely has a single measure that single sensing system can detect. As there are also cross-sensitivities of measurable parameters, Frost et al. (1997) proposed that systems need an integration of multiple data sources and historical and metadata to make sensible insights and predictions about animal health.

We also have the need for metadata (data about data) such as parity, age, number of live calves and days in milk, which need to be extracted from other data sources. For example, Lark et al. (1999) used a time series model of milk yield in early lactation to identify the development of ketosis, when metadata such as age and days in milk were added accuracy increased. In practice, human entered metadata needs curating to a standardised vocabulary of disease conditions before it can be searched with confidence (VirtualVet, 2021, Chapter 10 of this book).

Many systems do not produce data outputs that can be cross-related to each other. Some only offer a .pdf output, time stamps are not unified, and data columns in .csv files are not labelled. Data integration is not a technical problem; it is a commercial and social problem. It is expensive for companies to re-design software and there is a reluctance to open the systems to provide application protocol interfaces. To bring the major manufacturers to agree on a standardisation, an industry initiative is needed under the auspices of ISO or through a consortium approach such as ISO-BUS. A development platform is required that provides standardised published interfaces to permit data integration, but much software uses legacy code that predates these concepts of data integration.

Antanaitis et al. (2021) described the results of combining multiple data streams of behaviour and show that integrated data are available and capable of greatly improving cow health management. Fotea et al. (2020) built a bespoke collar system with an architecture designed for data integration and showed that it was possible to integrate data. However, they did find that 'there is no common or consensual standard to ensure a suitable interoperability between data and establish an easy collaboration/exchange between systems'. Any data input by humans, for example, a disease treatment, needs to be in a standardised format. The solution found by VirtualVet in Ireland was that when an incident was reported and added to a database of treatments, for example, by an image of the medicine bottle and the animal ID, a human operator would call the farmer to curate the data and identify the outcome in a standardised vocabulary.

Animal welfare

In recent years, consumers have taken an increasing interest in the quality of life enjoyed by dairy cows. The supply chain has responded by setting up monitoring programmes, usually based on periodic visits by an auditor, which are more a general assessment of the management conditions (cow density, cleanliness, air flow) than individual animal welfare assessments and incidence of disease causing pain and harm to the animal (mastitis and lameness). These are currently based on paper records of clinical judgements and treatments, which may not be complete or consistent.

These assessments can be high-tension events as a failed assessment may threaten the farms' future as a milk supplier. These are uncomfortable judgement calls; however, replacing the assessment by numerical systems has been criticised as setting up a binary process of a complex and dynamic multifactorial system. The continuous health monitor has successfully replaced these uncomfortable judgment calls on animal health by analysing data from multiple sources (Brouwer et al., 2015). However, predicting animal welfare using data sources was less successful (de Vries et al., 2013).

Chapa et al. (2020) reviewed sensor methods of monitoring cow and pig welfare with, for example, measuring lying time, which is an indicator of comfort. They concluded that there were already technologies for this on farm. These comfort data could be added to the health monitoring data to provide a welfare assessment. Automated data capture will at least record everything that the systems deem an incident, but we have to be aware of the risks of sensor and software errors that could generate many false positives (and negatives) and these could be an existential threat to the farm survival even when there are no real problems.

Conclusions

There is considerable activity in developing and launching new disease and welfare monitoring paradigms driven by the increasing scale of dairy operations and the market demand to improve animal health and welfare. The cost–benefit analysis of disease monitoring is not yet clear, but the demand by milk buyers for digital provenance data is creating a market, particularly for mastitis and lameness. Most of the new products are based on existing sensors such as tri-axial accelerometers and milk conductivity with more data analysis and integration. These approaches will have limitations as not all clinical diseases or welfare issues can be determined by change of behaviour, especially if it a slow subtle change. The most difficult time for a cow, at parturition, is still one of the least monitored, although a number of beneficial non-invasive technologies using accelerometers and dynamic image analysis are in development.

The greatest challenge is to redefine each disease condition in numerical quantitative terms rather than the qualitative clinical judgement of a veterinarian. Subclinical disease may be detectable, and prophylactic actions, such as changes of diet for metabolic disorders, palpation and increased milking frequency for mastitis, and foot trimming for lameness, can be taken well before a disease is visible to a clinician. That re-definition of disease into numbers from sensing systems would then be linked to standard operating procedures with a measurable outcome that can be identified and fed back to reinforce best practice.

New sensing systems are based on immunosensors targeted at biomarkers of the immune response and are likely to be deployed periodically to screen the whole herd on a given day rather than continuously. These new technical abilities detect health problems at clinical and sub-clinical levels. Subclinical detection probably needs more engagement from veterinary science to identify treatment pathways, as over-reliance on antibiotics has created a problem of antimicrobial resistance which has been identified as a global existential threat to humans and animals (WHO, 2021). Standard operating procedures to guide staff in an appropriate response are not yet widely published and financial models to guide decisions are not available.

The future of automated dairy cow health management will increasingly focus on data integration and software analysis. The speed of development of new technology has overtaken the ageing and possibly conservative veterinary profession. Quantitative models need to be developed and demonstrated within the training of veterinarians, so that they demand and use available farm data to improve diagnostics. The future of this area for engineers, veterinarians and dairy scientists is very interesting and will provide new diagnostic systems over the next decades.

References

Ackerley, E. (2014). *Assessment of a commercial pressure plate system (StepMetrix™) for the detection of lameness in dairy cattle, April 2014*. University of Aberystwyth. https://doi.org/10.13140/RG.2.2.28695.98720

Adams, R. S., Stout, D. L., Kradei, D. C., et al. (1978). Use and limitations of profiles in assessing health or nutritional status of dairy herds. *Journal of Dairy Science, 61*(1978), 1671–1678.

Alsaaod, M., Fadul, M., & Steiner, A. (2019). Automatic lameness detection in cattle. *The Veterinary Journal, 246,* 35–44. https://doi.org/10.1016/j.tvjl.2019.01.005. ISSN 1090-0233.

Anglart, D. (2014). *Automatic estimation of body weight and body condition score in dairy cows using 3D imaging technique.*

Antanaitis, R., Juozaitienė, V., Malašauskienė, D., Televičius, M., Urbutis, M., & Baumgartner, W. (2021). Relation of automated body condition scoring system and inline biomarkers (milk yield, β-hydroxybutyrate, lactate dehydrogenase and progesterone in milk) with cow's pregnancy success. *Sensors, 21*(4), 1414. https://doi.org/10.3390/s21041414, 1424-8220.

Atkinson, O. (2013). *A cross-sectional survey to investigate prevalence of and clinical indicators for Subacute Ruminal Acidosis (SARA) in lactating cows on UK dairy farms.* RCVS Diploma. https://knowledge.rcvs.org.uk/document-library/diplomas-dchp-13-1/.

Bewley, J. M., Peacock, A. M., Lewis, O., Boyce, R. E., Roberts, D. J., Coffey, M. P., Kenyon, S. J., & Schutz, M. M. (September 2008). Potential for estimation of body condition scores in dairy cattle from digital images. *Journal of Dairy Science, 91*(9), 3439–3453. https://doi.org/10.3168/jds.2007-0836. PMID: 18765602.

Bicalho, R. C., Cheong, S. H., Cramer, G., & Guard, C. L. (July 2007). Association between a visual and an automated locomotion score in lactating Holstein cows. *Journal of Dairy Science, 90*(7), 3294–3300. https://doi.org/10.3168/jds.2007-0076. PMID: 17582114.

Borchers, M. R., Chang, Y. M., Proudfoot, K., Wadsworth, B. A., Stone, A. E., & Bewley, J. M. (2017). Machine-learning-based calving prediction from activity, lying, and ruminating behaviors in dairy cattle. *Journal of Dairy Science, 100*(7), 5664–5674. https://doi.org/10.3168/jds.2016-11526. ISSN 0022-0302.

Britt, J. H. (2008). Oocyte development in cattle: Physiological and genetic aspects. *Revista Brasileira de Zootecnia, 37*(spe). https://doi.org/10.1590/S1516-35982008001300013. July 2008.

Brouwer, H., Stegeman, A., Straatsma, J. W., Hooijer, G. A., & van Schaik, G. (2015). The validity of a monitoring system based on routinely collected dairy cattle health data relative to a standardized herd check. *Preventive Veterinary Medicine, 122.* https://doi.org/10.1016/j.prevetmed.2015.09.009

Burfeind, O., Suthar, V. S., Voigtsberger, R., Bonk, S., & Heuwieser, W. (2011). Validity of prepartum changes in vaginal and rectal temperature to predict calving in dairy cows. *Journal of Dairy Science, 94*(10), 5053–5061. https://doi.org/10.3168/jds.2011-4484. ISSN 0022-0302.

CDC. (2019). *Centres for disease control.* https://www.cdc.gov/drugresistance/pdf/threats-report/2019-ar-threats-report-508.pdf.

Chapa, J. M., Maschat, K., Iwersen, M., Baumgartner, J., & Drillich, M. (2020). Accelerometer systems as tools for health and welfare assessment in cattle and pigs — a review. *Behavioural Processes, 181.* https://doi.org/10.1016/j.beproc.2020.104262, 104262, ISSN 0376-6357.

DeLaval. (2021). https://www.delaval.com/en-gb/discover-our-farm-solutions/delaval-delpro/precision-analytics/delaval-herd-navigator/2021.

De Vries, M., Engel, B., Den Uijl, I. E. M., van Schaik, G., Dijkstra, T., Boer, I. J. M., & Bokkers, E. A. M. (2013). Assessment time of the Welfare Quality protocol for dairy cattle. *Animal Welfare, 22,* 85–93. https://doi.org/10.7120/09627286.22.1.085

De Vries, A., & Marcondes, M. (2020). Review: Overview of factors affecting productive life-span of dairy cows. *Animal, 14*(S1), S155−S164. https://doi.org/10.1017/S1751731119003264

Drysdale, T. D., Mottram, T. T., & Cumming, D. R. S. (2008). WSN: Modelling the attenuation of radio signals by bovines. In *Proceedings of the American society of agricultural and biological engineers*. Paper number 084589, 2008 Providence, Rhode Island, June 29-July 2, 2008.

Dumont, B., Gonzales-Garcia, E., Thomas, M., Fortun-Lamothe, L., Ducrot, C., Dourmad, J. Y., & Tichit, M. (2014). Forty research issues for the redesign of animal production systems in the 21st century. *Animal, 8*(8), 1382−1393. https://doi.org/10.1017/S1751731114001281

Eckersall, P. D., & Bell, R. (2010). Acute phase proteins: Biomarkers of infection and inflammation in veterinary medicine. *The Veterinary Journal, 185*(1), 23−27. https://doi.org/10.1016/j.tvjl.2010.04.009. ISSN 1090-0233.

Fogsgaard, K. K., Løvendahl, P., Bennedsgaard, T. W., & Østergaard, S. (2015). Changes in milk yield, lactate dehydrogenase, milking frequency, and interquarter yield ratio persist for up to 8 weeks after antibiotic treatment of mastitis. *Journal of Dairy Science, 98*(11), 7686−7698. https://doi.org/10.3168/jds.2014-9204. ISSN 0022-0302.

Fotea, F. N., Roukha, A., Mahmoudia, S., & Mahmoudia S Aand Debauchea, O. (2020). Toward a big data knowledge-base management system for precision livestock farming. *Procedia Computer Science, 177*, 136−142. https://doi.org/10.1016/j.procs.2020.10.021. ISSN 1877-0509.

Frondelius, L., Nykänen, I., Lindeberg, H., Pastell, M., Koistinen, T., Palmio, A., & Ruuska, S. (2021). *Kuvaa Nautaa project: Thermal imaging in cattle health care*. Report of Savonia Institute. https://www.researchgate.net/publication/352380623_Kuvaa_Nautaa_-_lampokuvantamisen_hyodyntaminen_sorkkien_ja_poikimahalvausten_seurannassa.

Gasteiner, J., Fallast, M., Rosenkranz, S., Häusler, J., Schneider, K., & Guggenberger, T. (2009). Zum Einsatz einer intraruminalen pH-Datenmesseinheit mit kabelloser Datenübertragung bei Rindern unter verschiedenen Fütterungsbedingungen. *Veterinary Medicine Austria, 96*, 188−1194, 2009.

Gundelach, Y., Essmeyer, K., Teltscher, M. K., & Hoedemaker, M. (2008). Risk factors for perinatal mortality in dairy cattle: Cow and foetal factors, calving process. *Theriogenology, 71*(6), 901−909. https://doi.org/10.1016/j.theriogenology.2008.10.011. ISSN 0093-691X.

Halachmi, I., Klopčič, M., Polak, P., Roberts, D. J., & Bewley, J. M. (2013). Automatic assessment of dairy cattle body condition score using thermal imaging. *Computers and Electronics in Agriculture, 99*, 35−40. https://doi.org/10.1016/j.compag.2013.08.012

Hogeveen, H., Klaas, I. C., Dalen, G., HonigH, Zecconi A., Kelton, D. F., & Sánchez Mainar, M. (2021). Novel ways to use sensor data to improve mastitis management. *Journal of Dairy Science, 2021*. https://doi.org/10.3168/jds.2020-19097. ISSN 0022-0302.

Hovinen, M., Aisla, A.-M., & Pyörälä, S. (2006). Accuracy and reliability of mastitis detection with electrical conductivity and milk colour measurement in automatic milking. *Acta Agriculturae Scandinavica, Section A—Animal Sciences, 56*(3), 121−127. https://doi.org/10.1080/09064700701216888

Jonsson, N. N., Kleen, J. L., Wallace, R. J., Andonovic, I., Michie, C., Farish, M., Mitchell, M., Duthie, C. A., Jensen, D., & Denwood, M. J. (2019). Evaluation of reticuloruminal pH measurements from individual cattle: Sampling strategies for the assessment of herd status. *The Veterinary Journal, 243,* 26–32. https://doi.org/10.1016/j.tvjl.2018.11.006. ISSN 1090-0233.

Kamphuis, C., Frank, E., Burke, J. K., Verkerk, G. A., & Jago, J. G. (2013). Applying additive logistic regression to data derived from sensors monitoring behavioral and physiological characteristics of dairy cows to detect lameness. *Journal of Dairy Science, 96*(11), 7043–7053.

Keceli, A. S., Catal, C., Kaya, A., & Tekinerdogan, B. (2020). Development of a recurrent neural networks-based calving prediction model using activity and behavioral data. *Computers and Electronics in Agriculture, 170.* https://doi.org/10.1016/j.compag.2020.105285, 105285, ISSN 0168-1699.

Koczula, K. M., & Gallotta, A. (2016). Lateral flow assays. *Essays in Biochemistry, 60*(1), 111–120. https://doi.org/10.1042/EBC20150012, 2016.

Krieger, S., Oczak, M., Lidauer, L., Berger, A., Kickinger, F., Öhlschuster, M., Auer, W., Drillich, M., & Iwersen, M. (2019). An ear-attached accelerometer as an on-farm device to predict the onset of calving in dairy cows. *Biosystems Engineering, 184,* 190–199. https://doi.org/10.1016/j.biosystemseng.2019.06.011. ISSN 1537-5110.

Kumar, S., Singh, S. K., Singh, R., & Singh, A. K. (2017). Recognition of cattle using face images. In *Animal biometrics: Techniques and applications* (pp. 79–110). Springer Singapore. https://doi.org/10.1007/978-981-10-7956-6_3

Lark, M., Nielsen, B., & Mottram, T. T. F. (1999). A time series model of daily milk yields and its possible use for detection of a disease (ketosis). *Animal Science, 69*(3), 573–582. https://doi.org/10.1017/S1357729800051420

Leonardi, S., Marchesi, G., Tangorra, F. M., & Lazzari, M. (2013). Use of a proactive herd management system in a dairy farm of northern Italy: Technical and economic results. *Journal of Agricultural Engineering, 44*(s2). https://doi.org/10.4081/jae.2013.283

Mazeris, F. (2010). DeLaval herd navigator: Proactive herd management. In *Proceedings of first North American conference on precision dairy management* (pp. 26–27).

Mello, H., Bueno, P. R., & Mulatoa, M. (2020). Comparing glucose and urea enzymatic electrochemical and optical biosensors based on polyaniline thin films. *Analytical Methods, 34,* 2020.

Miedema, J. M. (2009). *Investigating the use of behavioural, accelerometer and heart rate measurements to predict calving in dairy cows.* PhD thesis). Edinburgh University.

Miedema, H. M., Cockram, M. S., Dwyer, C. M., & Macrae, A. I. (2011). Behavioural predictors of the start of normal and dystocic calving in dairy cows and heifers. *Applied Animal Behaviour Science, 132*(1–2), 14–19. https://doi.org/10.1016/j.applanim.2011.03.003. ISSN 0168-1591.

Moocall. (2021). https://www.moocall.com/calving/.

Morin, P. A., Chorfi, Y., DubucJ Roy, J. P., Santschi, D., & Dufour, S. (2017). Short communication: An observational study investigating inter-observer agreement for variation over time of body condition score in dairy cows. *Journal of Dairy Science, 100.* https://doi.org/10.3168/jds.2016-11872

Mottram, T. T. (1997). Automatic monitoring of the health and metabolic status of dairy cows. *Livestock Production Science, 48,* 209–217, 1997.

Mottram, T. T. (2011). *Monitoring of livestock.* GB patent 111396.6.

Mottram, T. T., Lowe, J., McGowan, M., & Phillips, N. (2008). Technical note: A wireless telemetric method of monitoring clinical acidosis in dairy cows. *Computers and Electronics in Agriculture, 64,* 45–48.

Mottram, Dobbelaar P., Schukken, Y., Hobbs, P. J., & Bartlett, P. N. (1999). An experiment to determine the feasibility of automatically detecting hyperketonaemia in dairy cows. *Livestock Production Science, 61*(1), 7−11. https://doi.org/10.1016/S0301-6226(99)00045-7. ISSN 0301-6226.

Neethirajan, S. (2017). Recent advances in wearable sensors for animal health management. *Sensing and Bio-Sensing Research, 12*, 15−29. https://doi.org/10.1016/j.sbsr.2016. 11.004. ISSN 2214-1804.

Ouellet, V., Vasseur, E., Heuwieser, W., Burfeind, O., Maldague, X., & Charbonneau, E. (2015). Evaluation of calving indicators measured by automated monitoring devices to predict the onset of calving in Holstein dairy cows. *Journal of Dairy Science, 99*(2), P1539−P1548. https://doi.org/10.3168/jds.2015-10057

Pastell, M., Hanninen, L., de Passille, A. M., & Rushen, J. (2010). Measures of weight distribution of dairy cows to detect lameness and the presence of hoof lesions. *Journal of Dairy Science, 93*, 954−960, 2010.

Pastell, M., Tiusanen, J., Hakojarvi, M., & Hanninen, L. (2009). A wireless accelerometer system with wavelet analysis for assessing lameness in cattle. *Biosystems Engineering, 104*, 545−551, 2009.

Post, C., Rietz, C., Büscher, W., & Müller, U. (2021). The importance of low daily risk for the prediction of treatment events of individual dairy cows with sensor systems. *Sensors, 21*, 1389 https://www.mdpi.com/1424-8220/21/4/1389.

Roe, E., Buller, H., & Bull, J. (2011). The performance of farm animal assessment. *Animal Welfare, 20*, 69−78. ISSN 0962-7286.

Rossi, R. S., Amarante, A. F., Correia, L. B. N., Guerra, S. T., Nobrega, D. B., Latosinski, G. S., Rossi, B. F., Rall, V. L. M., & Pantoja, J. C. F (2018). Diagnostic accuracy of Somaticell, California Mastitis Test, and microbiological examination of composite milk to detect Streptococcus agalactiae intramammary infections. *Journal of Dairy Science, 101*(11), 10220−10229. https://doi.org/10.3168/jds.2018-14753. ISSN 0022-0302.

Royal, G. D. (2021). https://www.gdanimalhealth.com/en/Animal-Health/MonitoringSurveillance.

Rutten, C. J., Velthuis, A. G. J., Steeneveld, W., & Hogeveen, H. (April 2013). Invited review: Sensors to support health management on dairy farms. *Journal of Dairy Science, 96*(4), 1928−1952. https://doi.org/10.3168/jds.2012-6107

Santman-Berends, I. M. G. A., Schukken, Y. H., & van Schaik, G. (2019). Quantifying calf mortality on dairy farms: Challenges and solutions. *Journal of Dairy Science, 102*, 6404−6417. https://doi.org/10.3168/jds.2019-16381

Schultz, L. H., & Megess, M. (1959). Milk test for ketosis in dairy cows. *Journal of Dairy Science, 42*, 705−710, 1959.

Singh, K., Molenaar, A. J., Swanson, K. M., Gudex, B., Arias, J. A., Erdman, R. A., & Stelwagen, K. (March 2012). Epigenetics: A possible role in acute and transgenerational regulation of dairy cow milk production. *Animal, 6*(3), 375−381. https://doi.org/10.1017/S1751731111002564. PMID: 22436216.

Sloth, K., Friggens, N. C., Løvendahl, P., Andersen, P. H., Jensen, J., & Ingvartsen, K. L. (2003). Potential for improving description of bovine udder health status by combined analysis of milk parameters. *Journal of Dairy Science, 86*(4), 1221−1232. https://doi.org/10.3168/jds.S0022-0302(03)73706-0. ISSN 0022-0302.

Song, X., Bokkers, E. A. M., van der Tol, P. P. J., Groot Koerkamp, P. W. G., & van Mourik, S. (2018). Automated body weight prediction of dairy cows using 3-dimensional vision. *Journal of Dairy Science, 101*(5), 4448−4459. https://doi.org/10.3168/jds.2017-13094. ISSN 0022-0302.

Song, X., van der Tol, P. P. J., Groot Koerkamp, P., & Bokkers, E. A. M. (2019). Automated assessment of reticulo-ruminal motility in dairy cows using 3-dimensional vision. *Journal of Dairy Science, 102*(10), P9076−P9081. https://doi.org/10.3168/jds.2019-16550. Oct 01, 2019.

Tasch, U., Erez, B., Lefcourt, A. M., & Varner, M. (2001). *US Patent Application 66992017, Method and apparatus for detecting lameness in animals.*

Turner, Claire, Knobloch, Henri, Richards, John, Richards, Peter, Mottram, Toby T. F., Marlin, David, & Chambers, Mark A. (2012). Development of a device for sampling cattle breath. *Biosystems Engineering, 112*(2), 75−81. https://doi.org/10.1016/j.biosystemseng.2012.03.001

Van Hertem, T., Maltz, E., Antler, A., Romanini, C. E., Viazzi, S., Bahr, C., Schlageter-Tello, A., Lokhorst, C., Berckmans, D., & Halachmi, I. (July 2013). Lameness detection based on multivariate continuous sensing of milk yield, rumination, and neck activity. *Journal of Dairy Science, 96*(7), 4286−4298. https://doi.org/10.3168/jds.2012-6188. Epub 2013 May 16. PMID: 23684042.

Viazzi, S., Bahr, C., Van Hertem, T., Schlageter Tello, A., Romanini, C. E. B., Halachmi, I., Lokhorst, C., & Berckmans, D. (2014). Comparison of a three-dimensional and two-dimensional camera system for automated measurement of back posture in dairy cows. *Computers and Electronics in Agriculture, 100*, 139−147. https://doi.org/10.1016/j.compag.2013.11.005

Villot, C., Meunier, B., Bodin, J., Martin, C., & Silberberg, M. (2018). Relative reticulo-rumen pH indicators for subacute ruminal acidosis detection in dairy cows. *Animal, 12*(3), 481−490. https://doi.org/10.1017/S1751731117001677. ISSN 1751-7311.

Voß, A. L., Fischer-Tenhagen, C., Bartel, A., & Heuwieser, W. (2021). Sensitivity and specificity of a tail-activity measuring device for calving prediction in dairy cattle. *Journal of Dairy Science, 104*(3), 3353−3363. https://doi.org/10.3168/jds.2020-19277. ISSN 0022-0302.

Wang, X., Gao, H., Gebremedhin, K. G., Schmidt Bjerg, B., Van Os, J., Tucker, C. B., & Zhang, G. (2018). A predictive model of equivalent temperature index for dairy cattle (ETIC). *Journal of Thermal Biology, 76*, 165−170. https://doi.org/10.1016/j.jtherbio.2018.07.013. ISSN 0306-4565.

Wikipedia. (2021). https://en.wikipedia.org/wiki/Glucose_meter.

World Health Organisation. (2021). https://www.who.int/news-room/fact-sheets/detail/antimicrobial-resistance.

Further reading

Schlecht, E., Hülsebusch, C., Mahler, F., & Becker, K. (2004). The use of differentially corrected global positioning system to monitor activities of cattle at pasture. *Applied Animal Behaviour Science, 85*(3−4), 185−202, 25 March 2004.

Voronin, V., Brayer, E., & Ben Menachem, U. (2006). *Method and device for detecting estrus.* Patent number: US7878149.

Whay, H. R., Main, D. C. J., Green, L. E., & Webster, A. J. F. (2004). Assessment of the welfare of dairy cattle using animal-based measures. *The Veterinary Record, 153*(7), 197−202.

WSAVA. (2021). https://wsava.org/global-guidelines/microchip-identification-guidelines/.

Fertility monitoring of cattle

7

Toby Trevor Fury Mottram

Digital Agritech Ltd, Kirkcaldy, United Kingdom

Introduction

Dairy cows are a major source of high-quality protein for humans. The modern global dairy industry has about 300 million cows in approximately 1 million herds that use industrial technology for feeding and milking. Milk consumption is growing steadily as it is a sound basis for nutrition for childhood development and milk provides a huge range of processed products such as butter, cheese, yogurts and kefir, and is the basis of many popular dishes. To produce milk, cows have to calve and many analyses have shown that the annual calving of cows is the most profitable method of maintaining a continuity of milk and heifer replacements. Since the introduction of artificial insemination in the 1960s, the onus of detecting when a cow is ready and fertile for insemination has moved from the bull to the herd's manager. Visual observation of oestrus has limited sensitivity and so increasingly technology has been introduced to solve the problem. In addition to detecting when the cow is ready to be inseminated, it is also vital to know when she is pregnant. This chapter describes why cow fertility management is such an important commercial objective, the underlying biological processes that engineers need to know and the various digital based approaches to solving this important problem.

For convenience, I will largely use UK data and spellings. The noun oestrus derives from Latin oestrus 'frenzy', originally from Greek οἶστρος oîstros 'gadfly'; American English has shortened this to estrus. The adjective oestrous (estrous) describes the cycle.

The UK dairy industry is fairly representative of the main systems of production and technology used in temperate climates; it contains a wide range of systems from outdoor grazing in the wetter, milder west, to intensive maize-based TMR systems in the drier South. Mean herd size is 150 and average cow yield per year is 8300 L. Herd sizes range from 60 cows up to a few herds greater than 2000 cows. The lactation length reduced from 337 days in 2010 to 319 days in 2017 in median herds within the sample. Conception rates improved, probably due to new behavioural monitoring technologies discussed in this chapter. The percentage of cows served by day 80 after calving increased by 14%–60% (70% in top 25% of herds). The percentage of cows pregnant by 100 days after calving has increased by 9%–35% (41% in top 25%). Conception rate has improved by 2%–34% (41% in top 25%).

Digital Agritechnology. https://doi.org/10.1016/B978-0-12-817634-4.00004-5

Submission rate improved by 11%–38% (49% in top 25%). Pregnancy rate has improved by 5%–14% (18% in top 25%) (Farmers Weekly, 2017). To summarise, only 1/3rd of cows were pregnant anywhere near that needed for an annual calving and the conception rate was half what it could be. That is the size of the problem.

This chapter is a written from an engineering point of view but has to describe the basic principles of the biological processes we are monitoring. A glossary in Appendix 1 gives definitions for some of the technical terms used. Biology in the field has to deal with whole living sentient creatures whose processes have been dissected using the powerful tools of reductionist microbiology. Thus, we know chemical signals by laboratory analysis of tissues and bodily fluids, but these are necessarily separated from the live animal and may not represent the full information necessary for recognition of an event such as oestrus. Our sensors are rarely directly accessing the measurands that give us the information we need and so we use proxy measures and models to relate them to oestrus. As there will also be measures of sensor accuracy (see Appendix 2 for definitions), these are another source of error. So we have multiple and accumulating sources of error in our sensing of whole animal biology; this is not precision engineering. That engineering systems are now capable of achieving detection of fertility events with greater than 97% sensitivity and specificity shows considerable progress over human detection of the same events in the order of 50%–70% and there is still room for improvement (Fig. 7.1).

Importance of fertility monitoring

Dairy farming is a commercial operation and thus the incentive to have well-managed fertility is largely financial. For optimum production efficiency, cows should calve every year for up to 15 lactations over 20 years. That very few cows

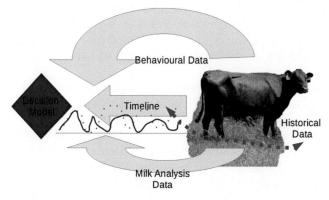

FIGURE 7.1

Using proxy measures (behavioural oestrus) and models we can improve the precision of the detecting the fertility status but risk multiplying errors in any part of the system.

achieve this is largely due to management failure. In fact, the mean age of leaving the herd is about 6 years of age with a typical 3.6 lactations over 4 years. The main cost comes from increased culling of animals deemed to be infertile. Then there is the loss of milk production from extended lactations and also the extra cost of multiple inseminations and veterinary interventions.

In an industry that has had many successes in improving efficiency (nutritional, labour, milk hygiene and disease control to name four), the ability to manage the fertility of the cow does not seem to have improved except in elite herds where O'Sullivan et al. (2020) showed that elite herds in Ireland (1.57) had fewer inseminations per pregnancy compared to the average (1.8). As humans are increasingly rare on farms, it means we must develop automated technologies that can improve fertility management to the highest level.

Giordano et al. (2011) examining the cost-effectiveness of reproductive technologies showed that in US dairies the economic gain increased as the reproductive efficiency improved and are still positive even at high reproductive performance. Reproductive improvement results in higher milk productivity and, therefore, higher milk income over feed cost, more calf sales and lower culling and breeding expenses.

Culling

After a female calf is born, the farmer will take a decision as to whether to keep her as a replacement. There are substantial costs in raising heifers and the aim of the farmer is to bring her into the herd at the start of the preferred calving season at about 2 years of age. She will make a margin in milk sales less feed cost in her first lactation to payback the rearing costs and begin lifetime profitability during her second lactation. The more lactations she has then has, the more profitable she is to the farm (Fig. 7.2).

Unfortunately, many cows are deemed infertile when the farm fails to detect their ovulations and inseminate them. The main reason for culling is failure to get in calf and this is compounded by the dismal statistic that many cows sent for slaughter are in fact pregnant. Thus, the problem is not just getting the cows inseminated but also knowing when they are pregnant.

Longevity

The natural life of a cow should be upwards of 15 years, but poor management particularly of fertility shortens this toward a mean age of culling at 6 years having produced maybe 40 tonnes of milk. This represents a huge economic loss to the farmers and is a waste of a valuable animal resource and a waste of environmental resource. The world record for lifetime milk yield of over 210 tonnes was set by a cow 'Smurf' of 15 years of age in 2012. Clearly, this cow had high genetic potential but was also managed very well. That most cows achieve less than 20% of the

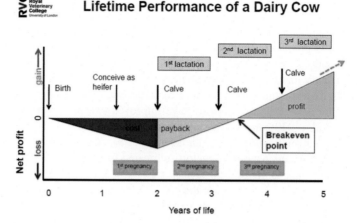

By kind permission of Claire Wathes Royal Veterinary College

FIGURE 7.2

The profitable life of a dairy cow usually begins in the second lactation. Few farms maximise the profits of longevity due to poor health and fertility management leading to many animals being culled before their break even point is reached.

Source: Copyright: 5M Books Ltd. (Wathes, 2021).

potential is the fault of systems of management. A key aim of improved fertility management is to give cows longer lives and enable farmers to make better choices to maximise profitability.

Climate change and dairy cow fertility

The dairy industry is increasingly criticised for methane emissions from enteric fermentation and slurry, and other pollutants such as ammonium, nitrates in water, and nitrous oxide emissions from fertiliser use. Methane is a global warming gas effectively 30 times more potent than carbon dioxide. Nitrous oxide emitted from fertilised land is also a potent global warming gas trapping long wavelength light. The best way to reduce these emissions is to improve efficiency since every adult cow emits up to 120 kg of methane every year. For example, if the replacement rate in a 500 cow herd improves from 30% to 20%, then there will 50 fewer heifers needed per year with a subsequent reduction of perhaps 5 tonnes of methane emitted per year. The societal and ethical benefits of reducing these emissions will in future probably be enhanced by incentive payments for low carbon systems adding another commercial benefit.

Biology of dairy reproduction

Although this chapter focuses on the engineering approaches to monitoring and improving cow fertility management, it is necessary to have an introduction to the biology to understand the different systems approaches.

The oestrous cycle

The oestrous cycle is a cyclical pattern of ovarian activity, which normally begins at 6—12 months of age and normally lasts between 18 and 24 days when the cow is not pregnant; it is controlled by a complex flood of hormonal signals to the reproductive organs. Progesterone is the main actor controlling the oestrus cycle by regulating the release of Follicle Stimulating Hormone (FSH) and Luteinising Hormone (LH).

The sequence of events associated with the ovarian cycle in dairy cows is well known and described in a number of standard texts such as Gordon (1996) and Peters and Ball (1994). After calving, the oestrous cycle is suppressed and restarts at about 15 days post-partum. Progesterone levels rise and then drop at the first ovulation of the lactation; thereafter, the oestrous cycle dominates until the cow conceives and progesterone then stays high during pregnancy.

The endocrine (hormonal) regulation of the oestrous cycle comes from a combination of positive and negative feedback; these hormones come from a variety of sources including the following:

- Hypothalamus — releases GnRH (gonadotropin-releasing hormone)
- Anterior pituitary gland — releases FSH (follicle stimulating hormone) and LH (luteinising hormone)
- Ovaries — releases P4 (progesterone), E2 (oestradiol), and inhibins
- Uterus — releases PGF (prostaglandin F2)

Progesterone is the most commonly measured hormone in relation to fertility and the estrous cycle firstly because it is found at a readily predictable and detectable level for long periods in comparison to other hormones such as oestrogens. Concentrations of progesterone in milk can be measured in nanograms per millilitre, whereas oestrogens are in picograms per mL (Fig. 7.3).

The typical progesterone profile of a healthy cow will follow this pattern:

- Luteal phase
  - period following ovulation, normally lasts 14—18 days
  - split further into met-oestrus and di-oestrus
  - following ovulation, the Corpus Luteum (CL) re-forms resulting in increasing progesterone concentration
  - increasing progesterone concentration following ovulation (met-oestrus) allows either the establishment of pregnancy or the resumption of normal oestrous cycling
  - during the rest of the luteal phase, progesterone concentration remains high
- Follicular phase
  - at the beginning of the follicular phase, the CL regresses causing circulating progesterone to decrease to basal levels
  - period between demise of CL (luteolysis) and ovulation, normally lasts 4—6 days
  - split further into pro-oestrus and oestrus

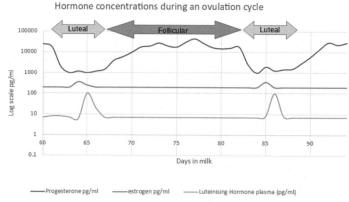

Hormone concentrations during an ovulation cycle

FIGURE 7.3

A schematic of the hormone concentrations around the key events of the oestrus cycle, the oestrogen and luteinising hormones spikes are of short duration and might fall between milking periods, whereas fall in progesterone is predictive and has a much larger change in concentration than other hormones.

○ defined by maturation and ovulation of ovulatory follicle, and the release of oocyte into the oviduct for fertilisation
○ Basal progesterone concentration (as well as other hormones) triggers a GnRH surge which allows the display of oestrus behaviours
○ progesterone concentration remains basal, and 10–14 h after oestrus, ovulation occurs marking the transition from follicular to luteal phase

The onset of behavioural, or standing, oestrus is stimulated by high levels of oestrogen which only occurs in the absence of progesterone. Hence, standing oestrus occurs soon after progesterone has dropped (due to the regression of the CL). The average duration of standing oestrus is 8.1 h; however, there is evidence that this is getting shorter and less intense.

Most cows resume ovarian activity within 15–45 days. However, this period can be extended due to problems with ovarian function, such as the following:

• Anovulatory anoestrus
• Prolonged luteal phase
• Follicular cysts
• Cessation of oestrous cysts

(Forde et al., 2010; Penn State, 2016b).

From a financial perspective, understanding when a cow has resumed cycling is very important. The optimum breeding time, in terms of economic gain, is the third oestrous cycle postpartum at approximately 80 days (Fig. 7.4).

Scientific knowledge of the sequence of events in the pre-ovulation period has been much assisted by the development of ultra-sound scanning which allowed

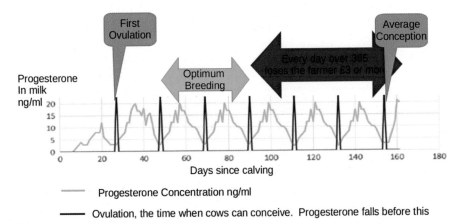

FIGURE 7.4

A timeline of progesterone concentration post-partum. To achieve annual calving, cows should be inseminated on the ovulation closest to 85 days post-partum.

Rajamahendran et al. (1989) to identify precisely the moment of ovulation in the oestrus cycle. Most cows display some form of oestrus activity before and after the peak in oestradiol although this can be mediated by many factors. Schofield (1988) describes diurnal peaks of oestrus behaviour early in the morning and late at night. However, he showed that standing to be mounted was not a reliable indicator of a cow being suitable to be served, with as many as 21% of ridden cows pregnant. Warren (1984) surveyed 35,000 cows in 255 herds and showed that 26% of inter-service intervals were greater than 48 days and that 19% of cows were served at an interval of 1−17 and 25−35 days suggesting that one or other of the observed oestrus events was not accompanied by ovulation whose periodicity has been established as very consistent. In the NMR survey of 500 herds in 2017, the inter-service interval of 18−24 days (heat detection) was only 36% (42% in top 25% of herds) compared to 55% in the Warren (1984) survey, showing that fertility management was declining on virtually every measure. This decline has not stopped but only slowed in recent years (Fig. 7.5).

The time of insemination in relation to the time of ovulation has an impact on the probability of conception and the sex of the foetus. Thus, systems for that predict the time of ovulation can have an advantage in providing control for the farmer and the management of the daily routine of insemination.

The challenge in reviewing the literature published about cow fertility management techniques since the 1960s is the changing nature of the farms and the cows on them. Many studies have been published with differing results and suffer from the large scale needed to identify small differences and non-reproducible results. Cow breeds have changed with the rise of the Holstein to global prominence with the belief that these animals are less fertile than other breeds such as Jerseys and the

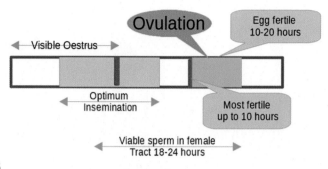

FIGURE 7.5

Time relationships of key reproductive events.

red breeds. Methods of feeding cows have changed radically and vary greatly in different geographic locations. There are seasonal variations in fertility (Norman et al., 2009) due to the light and temperature changes and these effects are rarely compensated for in experimental designs. Genetics for a whilst focused on milk yield, but with the introduction of type traiting, there has been a successful initiative to select for fertility as well. The farms have grown much larger and fewer humans are available to inspect and monitor cows. The wide use of slippery concrete floors and crowded pens inhibits oestrous behaviours (Penn State, 2016a). All these factors will have had an impact on the outcomes on farms and in experiments.

Post-insemination: pregnancy detection methods

Whilst the detection of oestrous and ovulation is key to correct timing of insemination and the subsequent conception, it is almost more important to detect the pregnancy status of the cow after insemination. After insemination, the chance of conception is rarely above 70% and even then the cow can 're-absorb' the foetus. After the first trimester, diseases and other factors can cause spontaneous abortion. The latter tends to be a highly visible event as the foetus is by then of substantial size and the distress of the cow is evident.

Once the cow has conceived, the progesterone concentration should rise and stay high throughout gestation. By testing on days when the concentration will go low if the cow has failed to conceive allows the farm to confirm pregnancy or intervene and inseminate again. Milk progesterone analysis has been conducted at 21 and 42 days after insemination since the 1970s by sending samples away for laboratory analysis. In recent years, on-farm test kits have been available which reduces the time between sampling and results and more recently inline milk analysis has become possible.

Pregnancy-associated glycoproteins (PAGs) can be measured in milk using ELISA tests from day 24 after insemination; this gives a positive indication of pregnancy which is not time dependent but comes too late to enable a second insemination before 100 days. Automating this test as a lateral flow immunosensor is becoming feasible and this would be a useful addition to inline progesterone monitoring as it confirms pregnancy throughout the first trimester rather than being time specific. There may be false positives if this is conducted too early in older cows as the glycoproteins can persist after calving.

Rectal ultrasound has been routinely used by farmers since the 1980s to listen for the foetal heartbeat, and with the introduction of ultrasonography (scanning) providing visual images of the foetus and ovaries, Fricke (2002) showed how this was a potent tool for pregnancy diagnosis with sensitivity and specificity if done at the right time. It can be used to confirm pregnancy from 60 days after insemination. The disadvantage of this technique is that it needs skilled labour and is a major disruption to the cow's routine and carries a small risk of injury and peritonitis.

Rectal palpation is probably still the most commonly used method of determining whether a cow is pregnant. It is a manual process which depends on the skill of the person for best results. It can be used 28–60 post-insemination and involves finding and feeling the status of the ovaries through the wall of the rectum. Paisley (1978) evaluated the impact of effectiveness of rectal palpation and identified that up to 6% of pregnancies could be lost after rectal palpation although whether this was a causal relationship is always going to be unclear. The main disadvantages to rectal palpation are the health and safety of the human and the disruption to the farm and cow's routine which has its effects on welfare and production. Milk progesterone analysis showed that rectal palpation was inaccurate as 36.5% of clinical diagnoses were incorrect (McLeod & Williams, 1991).

Systems of dairy cow fertility management
Natural service

Natural service has been used by farms since the domestication of cattle thousands of years ago. A bull can detect oestrus and serve a limited number of cattle (20–40 being usual). Where herds were bigger than 40–50 multiple bulls were needed and the problems of keeping them apart to prevent fighting become complex. The drawbacks were numerous; bulls are dangerous large animals and caused human fatalities and damage to farm equipment. Within 2 years, the bull would need to be rotated to avoid inseminating his daughters. Bulls can carry disease which cross infects within and between herds and primarily the use of a bull limits the potential for genetic gain covering maybe 1000 animals in a lifetime. Many farms keep a 'sweeper' beef bull to catch late breeders and a bull is also an excellent oestrus detector if kept close to the cow pens inducing behaviours and vocalising about them in the cows that might otherwise go undetected.

Conventional AI

Artificial insemination (AI) is the principal method used on commercial dairy farms to get cows pregnant. Foote, (2010) provides a history of this huge technical achievement. Semen is extracted from bulls at stud farms and then processed to extend it and allow it to be frozen. The bulls with highest genetic merit can provide millions of units of sperm which in turn can lead to hundreds of thousands of daughters worldwide. https://en.wikipedia.org/wiki/Toystory_(bull) AI also allows rare breeds to use a broader range of bulls than would otherwise be physically possible. Some studs use flow cytometers and other techniques to separate the sperm into those with XX and XY chromosomes which dominates the sexual orientation of the foetus. Thus, farms can inseminate (at higher cost) cows to have either female calves of high dairy genetic merit or bull calves for rearing for beef. The processing, freezing, and insemination process can all degrade the inherent fertility of the semen and this combined with need to get the timing right for optimal conception mean that accuracy of information is vital. A huge statistical framework exists to determine the efficiency of the inseminators and the fertility of the sperm from different bulls and the statistics reveal genetic traits that can inhibit conception. The biggest drawback of artificial insemination is the need to detect the oestrus event and this review is largely about the technologies to do this.

Synchronisation with hormones

Synchronisation or "the process" is a full intervention with reproductive hormones that overrides the natural oestrus cycle of the cow and creates a predictable ovulation event. It is widely used in the United States of America but is restricted in the EU to be used as veterinary intervention with cows that are deemed not fertile (Stevenson, 2016). The use of synchronisation requires a number of injections but does permit timed insemination; the conception rate is generally lower and the process has high costs for labour and drugs. There is a huge investment by drug companies in these processes and their marketing, but ethical and practical concerns remain Eriksson et al. (2018).[1]

Embryo Transfer is used by some breeders to implant fertilised eggs flushed from high genetic merit animals into surrogate dams of lower merit thus increasing the number of high genetic merit calves born. These techniques rely on excellent animal husbandry and monitoring of the fertility cycle.

Contracted systems (arm systems)

Contracting out fertility services to a third party is popular with large farms using total mixed rations and self-locking yoke feed fences. After cows are locked usually during a morning feed, a technician walks down behind the cows and inspects the tail paint (or chalk) previously applied to the tail head of every cow in the post-calving group. If the paint/chalk has been rubbed (indicating a standing to be served behaviour), the technician inseminates the cow on the basis that within the past 24 h she has started oestrus and is ready to inseminate.

---

[1] https://dairy-cattle.extension.org/dairy-herd-synchronization-programs/.

This system is very dependent on human skill and the success rate of the different technicians can be assessed by two overlapping criteria. Submissions — meaning the number of animals that the technician inseminates — indicate a measure of oestrus detection sensitivity. The conception rate will be the number of cows served who become pregnant and can be a measure of specificity. The technician with a high submission rate may more likely to be successful at getting more cows pregnant even though conception rate will be lower (as more cows not ready for insemination are presented).

Digital systems for fertility management
Mount detectors

Cows commonly exhibit an increase in activity level with certain characteristic behaviours in association with oestrus, for example, mounting other cows, or allowing themselves to be mounted and in general more walking activity. The amount of activity is impacted by the environment and slippery concrete surfaces will inhibit these activities (Britt et al., 1986). Van Vliet and Van Eerdenburg (1996) discuss the relative significance of each of these behaviours as indicators of oestrus. The most simple activity-related device is the heat mount detector, Williamson et al. (1972). A plastic phial containing dye is stuck to the tailhead. When the cow is mounted, the dye is squeezed from the phial and marks the cow. Gwazdauskas et al. (1990) report significant differences between oestrus detection rates by observation and by two type of rump marker. They also report a high (40%) loss rate of devices. Stevenson et al. (1996) report the use of an electronic, radiotelemetric, pressure-sensitive, rump-mounted device, which was designed to be activated by the weight of a mounting animal. The sensor sent a signal by radio to a computer which recorded the mounted animal's identification, date, time and duration of the mounting. These systems do not seem to be available in the market.

Early Oestrus Technology

Collars (and pedometers) have been used to detect oestrus for many years measuring the increase in activity above the base line of normal activity. The early versions used tilt switches based on a metal ball or mercury moving in a switch tube. When contact was made by the switch due to cow movements, a counter was incremented; the number of counts would then either trigger a continuous radio transmission as in the Alpro or the number of counts would be transmitted when the cow neared the antenna. In recent years, microchip accelerometers have displaced the old switch mechanisms and this opens new potential for monitoring cow health and behaviour. The amplitude of accelerations measured is used instead of the mechanical devices and software either on the collar or at the base station calculates an index. Cows move for many reasons and this can be filtered by software, for example, the Heattime system based on the Voronin et al. (2005) patent discards the data associated with feeding, as this improves the signal to noise ratio. In one implementation, the eating activity is measured with an acoustic device.

Pedometers

Pedometers are electronic devices which are strapped to a cow's leg, or hung around the cow's neck to register movement. There are various designs of pedometers, until about 2000 they typically contained an electrical switch which opened or closed when the device is moved. Each activation of the switch added to a total stored by an electronic counter. The total count was transferred to a base station by radio-telemetry. Since the number of movements counted by a pedometer has no absolute significance, it is necessary to establish a comparative technique in which a count is compared to a baseline count which would be expected from the cow if it were not exhibiting oestrus. If a count exceeds the baseline by some predetermined multiplier, the cow is deemed to be in oestrus. The main variables involved in this process are therefore how frequently the total counts are transferred to the base station (this is often twice a day at milking), and the multiplier that is used to set the oestrus alerts threshold. The multiplier can vary due to the surface the cows are walking on (Penn-State, 2016a).

There have been many studies aimed at assessing the effectiveness of pedometers. Peter and Bosu (1986) carried out a trial with 47 cows at pasture. Pedometer readings were taken twice a day at milking. No details are given of the alert threshold that was used, but 76% of ovulations were detected by pedometer, compared to 35% by herdsman observation (30 min twice a day). All of the cases that were detected by the herdsman were also detected by the pedometers. The reference method for ovulation detection was measurement of blood progesterone concentration.

The use of unstructured learning (machine learning) was tried in an experiment by Shahriar et al. (2016) to use collar accelerometer data. The number of oestrus events measured was 38 and they were attacking the difficult problem of monitoring grazing animals where activity data is confounded due to the travel of the animals to the pasture and back (Fig. 7.6).

Since the advent of cheap tri-axial accelerometers and digital signal processor chips in the 2000s, the collar has been shown as a tool that can not only detect oestrus but also lying behaviour, lameness, location and with the collection of audio data rumination and eating activities. These collars have been used widely and models are slowly emerging as to how the data can be integrated into a wholistic management system. Commercial activity has probably been more advanced than research reports with large developments in largely secret algorithms to extract information from the data. The algorithms have to cope with movement of the collar around the cows neck, activities such as rubbing and grooming and a generally black box approach is used to extract the gross change in activity associated with oestrus. The Voronin et al. (2005) invention provided a method and device for detecting oestrus in animals by sensing along with time the motion of the animal and identifying

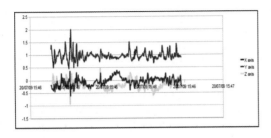 

FIGURE 7.6

Plot of three axis data from a cow collar; algorithms convert these accelerations into movements from which can be deduced behaviours associated with oestrus.

when the sensed motion is not related to eating periods of the animal. Kamphuis et al. (2013) had similar results to those with pedometers and achieved 76.9% SN, 99.4% SP, and 82.4% positive predictive value (PPV), while activity only collars achieved 62.4% SN, 99.3% SP, and 76.6% PPV all in comparison with progesterone analysis of milk as the gold standard. This and over 50 patents in this area show each attempting improvement on the basic concept. There has been major activity in new company formation selling solutions to this age old problem as the market expands; they are focused largely on data download (for example, by drone) and data aggregation and presentation via apps. Some companies such as SmartBow have focused on ear tag behavioural systems that include standing oestrous. Smaxtec uses a bolus system which has the benefits of security (eartags and collars are easy lost, damaged or stolen) as boluses are retained in the reticulum of the rumen and are thus there for the life of the cow. Since the rumen is a very active zone with continuous peristaltic muscular movements, considerable modelling is required to extract whole body movements from the noise.

The conclusion is that behaviour can never achieve 100% specificity of oestrus and thus ovulation and cannot detect pregnancy. There may be improvements in reliability, in terms of registering movement and transmitting the data to the base station. There is also a possibility that better use could be made of movement data if, rather than being taken as the sole indicator of ovulation, it were combined with data from other sources. A major issue is the difference in behaviours seen between cows (some will move more than others), as well as false positives and negatives (i.e., increase in movement for another reason or no increase in movement but cow is in oestrus). Other issues that arise with behavioural measurements are the battery life which is typically listed as several years, but as many devices get lost or damaged, the devices are typically replaced in less than 3 years.

Another potential technology for measuring oestrus is video analysis with cameras recording and analysing behaviour. Considerable research effort is being directed at the use of computers to analyse video images of moving objects for human recognition and behavioural monitoring. McFarlane and Schofield (1995), for example, reported automatically tracking the movements of pigs in images from an overhead camera. Tillett et al. (1997) have developed a means of tracking postural changes of pigs such as bending and head nodding, as well as position and rotation, again using an overhead camera. Bruyere et al. (2012) used four cameras to flag up behaviour in 35 cows with static images and then spending a mean of 20 min daily examining images. This system provided 80% detection of oestrus compared to 68.6% of the control manual observation. There is considerable engineering required to convert these concepts into full automation to detect animal behaviour and it would also need to be able to identify the active animals from the images. There are however some significant fundamental technical problems, related to the complexity and variability of the scenes to be analysed, that would have to be solved before a reliable, automatic, video analysis-based oestrus detection system could be developed. The installation of cameras and computers need not be expensive to install on a per cow basis as it avoids the cost of individual devices, but there seems to be little current development of this concept for dairy cows. Visual analysis technology is proceeding rapidly in other areas for human detection and behavioural analysis and similar techniques could then be applied to dairy cow management.

Pederson and Pedersen (1995) report the use of passive infrared detectors (PIDs) to measure animal activity. These were commercially available devices intended for security alarms. Using a laboratory rig with a moving heat source as a simulated animal, it was found that the output signal from a PID could be processed to provide a voltage which was proportional to the velocity of the simulated animal and the difference between the temperatures of the simulated animal and the environment. However, there was a non-linear response to the addition of a second heat source and this approach seems to have been abandoned.

Olfactory sensing

The natural method of oestrus detection does not involve any human agency and we assume that the bull relies on a combination of senses, olfactory, visual and auditory. Kiddy et al. (1984) trained dogs which were used to identify different bottles containing suitable body compounds of dairy cows and concluded that odours specific to oestrus were distributed throughout the body. Blazquez et al. (1988) showed that pheromonal odour secreted from the perineal glands near the vagina was the determinant of bull behaviour towards the cows. Klemm and his collaborators (1987, 1995, 1994) developed an understanding of the odours secreted in vaginal mucus. They identified acetaldehyde as a compound associated with oestrus, but the identity of the pheromone has still to be determined.

Attempts to develop an electronic nose gave results indicating that an array of tin oxide sensors could discriminate between oestrus and di-oestrus from the odour of vaginal swabs but not from air samples taken from the surface of the cow, but subsequent studies by the author did not produce encouraging results. The principle compound found was acetaldehyde which just rises in concentration with the amount of mucus being from the vagina or mouth (Mottram, Hart, et al., 2000).

Agscent (2020) has recently developed a handheld breath sampling device to detect early pregnancy in livestock. A sampling device is used collect a breath sample from a restrained cow approximately 60—80 days after insemination. The sensor then detects changes in various biomarkers which occur through the course of pregnancy. Agscent have a provisional patent application and are currently validating their proof of concept trials. They claim to be able to detect pregnancy at 60 days post-insemination and in around 30 s. In the future, they are aiming to refine the process and integrate the sampling device and nanosensor into a single unit. It is unclear which biomarkers they are measuring so it is difficult to say if the technology will even work in principle. The focus appears to be range beef cattle where animal handling is rare, and as the technology is not inline, requiring the cows to be restrained causes a change of routine and unnecessary stress during pregnancy.

Milk temperature

It is well known that a cow's body temperature rises at oestrus. There have been many attempts to base oestrus detection systems on these changes. Some methods have been invasive, using devices such as intravaginal temperature transmitters, but they are too invasive to be considered here. An alternative, non-invasive approach is to measure the milk temperature in the claw piece or short milk tube of the milking system. Maatje et al. (1987) carried out two experiments; one with 28 housed cows, and one with 20 cows that were grazing during the day and housed at night. Oestrus was assumed to be associated with a significant (twice the standard deviation of the temperature during the previous 5 days) rise in milk temperature, compared to the average temperature over the five previous days. This produced an oestrus detection sensitivity of 74%, with a false rate of 8%, using milk progesterone as the reference technique. It was concluded that the relationship between oestrus and milk temperature was subject to confounding factors such as ambient temperature, and that some cases of oestrus were not accompanied a detectable rise in milk temperature.

McArthur et al. (1992) also examined the reliability of this method. They made measurements under experimental, controlled conditions, and under commercial conditions. The milk temperature of the two cows studied under controlled conditions rose by about 0.4°C on the day when behavioural oestrus was observed. On a commercial farm, milk temperatures were measured for 18 cows which exhibited a total of 34 periods of oestrus. Setting a threshold of 0.3°C elevation in temperature over the average for the previous 5 days resulted in an oestrus detection sensitivity of 50% (i.e., 50% of actual oestrus incidents were detected), with an associated false rate of 81% (i.e., 81% of oestrus indications were false). Increasing the threshold elevation to 0.6°C resulted in a reduction in the false rate to 65%, but a reduction in sensitivity to 32%.

McArthur et al. (1992) concluded that the duration of temperature change (under 9 h) period and ambient temperature changes led to a clear conclusion that the detection of oestrus based on twice daily measurement of milk temperature is not reliable.

Milk yield

A number of authors (Blanchard et al., 1987; Lewis, 1984; Schofield et al., 1991) have suggested that the continuous monitoring of milk yield and other parameters will indicate oestrus.' However, this early application of big data analysis did not detect patterns that were sufficiently specific to be of value to detect oestrus.

Skin temperature

Hurnik et al. (1985) investigated the possibility of using thermal infrared scanning of the body surfaces of a cow to detect temperature changes related to ovulation. The study was conducted using 27 cows, housed in tie stalls. A thermal imaging device, comprising a temperature sensing camera and a video display, was used to take images of the gluteal region of the cow, including the anal and vulval areas, the posterior zone of the udder attachment and the two posterior lobes of the udder. According to the specification of the imaging system, it was possible to resolve differences of 0.2°C. The images were analysed by measuring the total area of the cow that was enclosed a given (37°C was the value used) isotherm. Using a criterion based on a given percentage increase in this area, the oestrus detection sensitivity was about 80% (using milk progesterone as the reference technique), but this was associated with a false rate of 33%. The overall conclusion was that high frequency of false positives and false negatives meant that the technique was not suitable for routine oestrus detection. There have been huge improvements in thermal imaging equipment, but the difficulty of eliminating or accounting for non-oestrus-related temperature variation caused by factors such as environmental temperature variation, and moisture on the skin, remains.

Combined measurements

Several researchers have suggested that whilst measuring one particular variable may not yield an adequate prediction of ovulation, accuracy might be improved by measuring more than one variable, and basing the prediction on a combination of the measurements. Maatje et al. (1997) report a multivariate oestrus detection model which bases detection on a combination of activity (as measured by pedometer), milk temperature, yield and feed intake. Using data from two experimental farms, which included over 500 cases of oestrus, they achieved a detection sensitivity of 87% with a specifity of 97%. This represented an improvement in sensitivity, with equal specificity compared to results obtained using activity alone. Mitchell et al. (1996) have carried out a preliminary investigation of the possibility of combining milk yield and data on milking order to detect oestrus. These variables were chosen because they were thought to be the most appropriate for New Zealand dairy herds. The proposition was that, since at oestrus milk volumes sometimes fall and then rise at the next milking and the order in which cows present themselves to be milked changes, it may be possible to use a computer to recognise characteristic patterns in the data. Two different machine learning procedures (C4.5 and FOIL) were tried on a year's data from a herd of 130 cows. The best result that was achieved was a sensitivity of 69%, with an associated false rate of 74%. The overall conclusion was that some fundamental questions regarding the nature of cow performance variations at and around oestrus remain to be answered. It was also suggested that performance could be improved by including more monitored variables. The approach of using big data and multiple variables would require huge amounts of learning matched to accurate metadata to enable algorithms to be honed; this has not happened yet.

Measuring hormones in milk
Progesterone assay

Laboratory tests for progesterone in serum and later in milk were developed in the 1970 and 1980s. On-farm test kits to detect changes in the level of progesterone secreted in milk have been available for some years. A sample of milk is taken and analysed once per day firstly to identify that oestrus cycling has begun and then to identify the drop in progesterone which precedes ovulation by approximately 48 h. Nebel (1988) reviewed the development of the immunosensing tests for progesterone. The principle of these tests is that an antibody to progesterone is attached to a surface of a test well during manufacture. The farmer or veterinarian then adds milk. Progesterone in the milk then attaches to the antibody coated on the surface. Various other compounds are added and rinsed out and a colour change is observed to identify the amount of progesterone present in the milk. Although in laboratory terms this is a simple test, it does require a small amount of skill, some space and correct timing to get consistent results. Analysis of progesterone levels in milk can not only be used to monitor the stage of the oestrous cycle but also to detect pregnancy and to identify early ovarian disorders (Ball & Jackson, 1979)'. Since taking and analysing milk samples away from the milking parlour is labour intensive, strategies have been needed to minimise the number of samples needed to improve fertility. Experimental work during the eighties established a sampling protocol and showed that inseminating cows on the basis of progesterone profiles using laboratory analysis of samples could achieve significant improvements in fertility (Foulkes et al., 1982; Foulkes & Goodey, 1988). The protocol used by McLeod et al. (1991) showed in an experiment that 99% of 88 ovulations were correctly identified using on-farm progesterone kits compared with 78% in a control group monitored conventionally. Milk samples were taken three times a week starting 25 d postpartum. Once an ovulation had been detected by a fall in progesterone concentration to below 4 ng/mL and a subsequent rise to more than 7 ng/mL, sampling was suspended for 15 days. Sampling resumed on alternate days until a fall in progesterone indicated the onset of oestrus. The cows could then be inseminated 48 h following the fall in progesterone. Sampling continued so as to determine whether oestrus had been correctly identified or whether the cow had conceived.

The sampling and analysis protocol used by McLeod and Williams (1991) was also used to detect ovarian malfunction. A study of over 500 cows in a controlled trial suggested that alternate day progesterone profiles were a better method of analysing ovarian malfunction than rectal palpation. This study confirmed other results and showed that there was little incidence of ovarian dysfunction and that the principal cause of extended calving intervals was a failure to detect oestrus. Cows were diagnosed as anoestrus — when progesterone was below 4 ng/mL for 30 d postpartum. Ovulation was deemed to have occurred if progesterone was below 4 ng/mL followed by 5 days of progesterone rising above 4 ng/mL, with at least one sample greater than 7 ng/mL. Normal cycling was detected by an increase of progesterone which remained high for more than 5 days and less than 18 days. Insemination was assumed to be correctly timed if progesterone increased in the following 2—6 days. Conception was identified if ovulation and correctly timed insemination coincided and that progesterone remained above 4 ng/mL for more than 20 days. If progesterone was greater than 4 ng/mL for more than 30 days after conception, then pregnancy was assumed to be established. By contrast, embryo loss was assumed to have occurred if progesterone was below 4 ng/mL following conception. A luteal cyst could be diagnosed if progesterone was above 4 ng/mL for more than 20 d following an ovulation where no insemination was made. A follicular cyst could be diagnosed if progesterone was less than 4 ng/mL for more than 8 days after a high level. At 42 days post-partum, 92% of cows were cycling normally which reinforces the vie w that identification of the oestrus cycle is the problem rather than fertility as such. Williams and Esslemont (1993) describe the effectiveness in field conditions of the improvement in fertility achieved by using the MOIRA (management of insemination based on routine analysis) which is based on the routine use of progesterone assays. Ovulation detection rates of 98% were achieved.

The milk tests to measure progesterone are still available, but the manual labour required has meant that very few farmers use this accurate technique. What has been needed is full automation of inline milk analysis.

Biochemistry of hormone analysis

The ELISA test has become a standard in biomedicine and many textbooks are available for those interested. Similarly, the newer technique of lateral flow immunosensing is a whole subject in itself, so a short engineering interface description is necessary.

The term biosensors is loosely used to describe a number of different devices that can measure specific, sometimes complex, biological molecules by a change in electrical or opto-electronic signal. A biosensor system consists of a sensor, a system to interrogate it and a computer to convert the signal into a format to be added to a database and displayed to an operator. The sensing methods are usually based on a monoclonal antibody which is specific to the compound being detected. A review of systems suitable for agricultural applications was conducted by Velasco-Garcia and Mottram (2003).

A commercially available biosensor of relevance to this study is the human pregnancy test system sold over the counter. The sensing medium is a disposable screen printed sensor with monoclonal antibodies to a metabolite of oestradiol and luteinising hormone. The electrical circuit for this type of sensor is a relatively simple optical transducer and is contained in purse-sized container along with the microprocessor and display and a small battery pack. Monitoring hormones in milk has slightly different application problems.

Milk as it is drawn from the udder is not homogeneous. The initial foremilk is often contaminated through first contact with the ambient environment. The milk fat concentration rises through milking, and as the progesterone is bound up in lipid bundles, its concentration also changes during milking. Samples should therefore either be taken from the mixed milk at the end of milking or from a point during milking when the concentration is generally representative.

Progesterone is a very small steroid molecule of about 254 Da and it is in a heterogenous aqueous solution full of lipid bundles and proteins of much larger size, the opacity of milk and the molecular structures make spectroscopic analytical methods such as infra-red analysis and mass spectrometry unsuitable for adaptation for inline analysis. The large size of the lipids and proteins induces steric hindrance that in the lab is overcome with reagents. An antibody-based approach is needed to capture the free progesterone in a physical and electrical bond which then changes a measurable characteristic of the binding surface.

Koelsch et al. (1994) reported a method of detecting progesterone with a quartz crystal micro-balance device. The operating principle was that a crystal with a known natural frequency of oscillation was coated with an antibody which would bind to progesterone when exposed to it. Since the mass of the progesterone binding to the antibody would change the frequency at which the crystal oscillated, the degree of binding could be determined by measuring a change in oscillation. The device was dipped into a solution of progesterone and then exposed to air and the change in oscillation monitored. The crystal would have to be dipped into milk and then removed so as to allow the crystal to vibrate at its natural frequency, which would require some fine scale engineering. Even if these problems could be overcome as was pointed out by Claycomb and Delwiche (1998), realistic levels of progesterone would only produce a change in mass of 0.4% very close to the noise level of the microelectronics. Claycomb and Delwiche (1998) proposed an alternative solution by developing in effect an automated ELISA test with a biosensor that was developed for on-line measurement of progesterone in bovine milk and detection of oestrus. The biosensor used an enzyme immunoassay format for molecular recognition, which was developed to run in approximately 8 minutes. The sensor was designed to operate on-line in a dairy parlour using microinjection pumps and valves for fluid transport, fibre optics and photodiodes for light measurement and a control computer for sequencing. Calibration showed a dynamic response between 0.1 and 5 ng/mL progesterone in milk. The reusability of the test well was evaluated.

Thiocyanate (0.5M, pH 5.1) quickly regenerated the antibody surface whilst maintaining antibody activity for 15−20 cycles, but noise from the residual enzyme limited reusability.

Well cow

Mottram, Lark, et al. (2000) described an automated system to measure milk progesterone inline using a disposable voltametric sensor. This system was then developed and deployed at the Royal Veterinary College, Potters Bar, UK, in 2003 and a video of it in operation made by the Well Cow company is set up to commercialise; it is available on YouTube https://www.youtube.com/watch?v=KKlS5vceAGM.

On arrival in the stall, the animal was identified by transponder. The system interrogated the database to determine if the cow needed to be sampled, if she was, the system waited for a signal to indicate when the cow has started milking. Ninety seconds after the start of milking, the sampler pulled a milk sample from the sample point to a collector. The collector then held the sample until the analyser was ready. Up to 12 samples could be held at a constant 37°C until the analyser was ready. The analyser took about 15 min to complete the assay, the result of which was added to the database for each animal. The sampler, sample lines and sample points were washed as part of the milking system circulation cleaning in place. Unfortunately funding did not take this prototype to commercialisation.

Herd navigator

Based on a patent filed in Carlsen et al. (2004), Foss Electric and de Laval developed the Herd Navigator system which combines automated sampling and five sensing systems including progesterone in milk. It has a similar structure to the Well Cow system although it appears to only have been fitted to robotic milkers. A sampling pipe leads from the milk lines to a box with a controlled environment containing effectively a laboratory style dispensing head which processes and applies milk to a sensor and appropriate reagents and sends an analog value of progesterone to the database. It also has the capability to monitor nutritional and mastitis markers although these are not seen as cost effective and rarely implemented in practice. Blom and Ridder (2010) reported that heat detection rates of 95%−97% were being achieved by Herd Navigator on three farms in Denmark with the number of days open reduced by 20 days in the first year of operation. Pregnancy rates also increased significantly in a very short time: up to 50%. Herd Navigator automatically measures the level of progesterone in milk, software indicates insemination time, lists animals for final pregnancy confirmation, indicates early abortion and lists the cows with risk for cysts and prolonged anoestrus. Vreeburg (2010) reported heat detection rates of two farms in the Netherlands using Herd Navigator as 94% and 99%. Pregnancy rates of 42% and 46% were reported. Mazeris (2010) claimed that a number of Herd Navigator farmers have stopped performing manual pregnancy tests and saved between €250 and €350 per cow per year. Herd Navigator consists of a large analyser boxed with controlled environment and complex plumbing system to bring the milk samples to the analyser from the short milk tubes. It was designed to fit into parlours with a small number of milking points when newly installed and would be difficult to retrofit into existing parlours, particularly large systems that are working continuously (Fig. 7.7).

Milkalyser

The author must here declare an interest as Milkalyser is the name he gave a system he invented and developed from 2015 onwards, filing a patent (Mottram, 2016) and taking this system through proof of concept, raising funding, demonstrating the technology and selling the company in 2020. The Milkalyser system was developed to address the main problems of the earlier Well Cow and Herd Navigator systems. It was designed to fit directly to the milk line from each cow in any conventional parlour or robotic milking system. The system was designed to be simple to fit in a few minutes using wireless data connection and single use lateral flow immunosensors packed in a cassette. It requires only a connection to the milk tube and a 24V power supply.

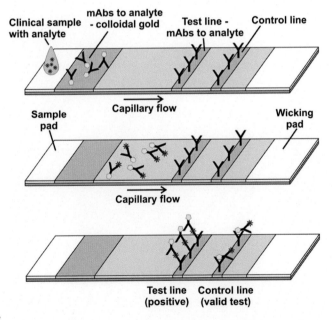

FIGURE 7.7

A packaged strip of lateral flow immunoassays with desiccant pill for the Milkalyser system of automated progesterone analysis. The machine punctured the packaging and injected 200 µL of milk onto the test zone and read the colour change with a Raspberry Pi camera.

The active component of the system was the Milkalyser Unit. One such unit was fitted to each point of the milking parlour. The unit drew a sample from the milk tube on demand and determined the high or low progesterone concentration with a single-use colorimetric sensor housed in a cassette (Fig. 7.8).

The units of an installation are controlled by a single Milkalyser Hub, which had a regular but not necessarily continuous internet connection to a cloud server, the Milkalyser Core. A connection to cow identification systems and milk flow monitoring was provided by the Milkalyser interface device, which was specialised to the messages flowing at each parlour installation.

When a cow was identified as entering a stall, the Milkalyser Hub determined whether the cow was to be sampled. This decision was governed by existing information on the animal and the level of accuracy desired by the user. If a cow was to be

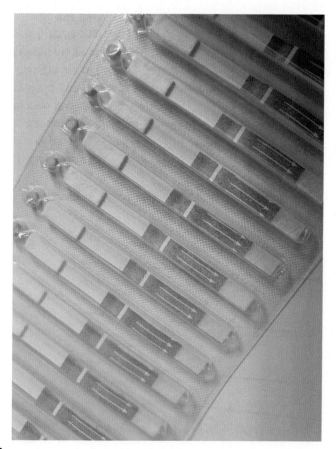

FIGURE 7.8

Milkalyser units attached on a rotary milking parlour; the requirement is for simple plug in and Wi-Fi connections.

sampled, the unit at the relevant stall was put at readiness. The cassette of 25 sensors in which each sensor was held in a self-contained air tight pocket was advanced into position. At a specified time into milking a sample was drawn from the milk line and applied to a sensor by a needle piercing the surface film of the sensor pocket. The microlitre quantity of milk would then rise through the chemical matrix of the lateral flow immunosensor setting off a colour reaction of a control and measurement strip at the top of the sensor. After allowing time for sensor development, an image of the sensor was captured. The control line indicated that a milk sample had been analysed and the measurement line appeared if the concentration of progesterone in the milk was below 4 ng/mL. The sample result was communicated to the Milkalyser Hub, which connected to the MC to update cow records and push action lists to the farmer, AI technician, etc. The Milkalyser effectively measures the mark to space ratio of the progesterone concentration. On a falling edge from mark to space during the cycling period, the indication is sent that the cow will ovulate in a number of hours and the decision to time the insemination can be taken. The sampling protocol can be manipulated in software to minimise the number of analyses for cost saving or increase the sampling frequency for problematic or more valuable animals. In the event of a doubtful result, the sampling and analysis can be repeated at the next milking. Post-insemination, the sampling can be activated for each day from 17 days to look for the transition from mark to space and either confirm conception or initiate insemination.

The Milkalyser would be ideally deployed on a robotic milker which provides a host of signals about cow identity, milk flow and wash cycles, etc. The Milkalyser company became a wholly owned subsidiary of Lely robotics in early 2020.

RePro

In 2019, the de Laval RePro system was launched, technical details are limited but we can presume that the system is protected by a patent (Rasmussen & Carlsen, 2017) which describes a system similar in concept to the Milkalyser. The principal difference is that the RePro deploys its sensors as a tape and the protecting tape is peeled away to allow milk to be applied to a "dry stick" that gives a colorimetric signal captured by a camera. The tape system also allows for different sensors to be deployed although this has not been reported independently.

Milkalyser and RePro are revolutionary tools which on their own will substantially improve ovulation prediction and pregnancy detection; however, monitoring progesterone can provide numerous other benefits. As well as detecting oestrus, mathematical models can be used interpret progesterone profiles and gain more information regarding the cow's health and reproductive status. A growing body of research has indicated that using progesterone profiles it may be possible to identify cows which may be suffering from reproductive disorders, are not pregnant following insemination, and those which may have an abnormal pregnancy. This information can then be used as a basis for further veterinary investigations.

Where multiple sensors to measure different analytes can be deployed, it will be possible to add other lateral flow immunosensors. One analysis of proven value for later pregnancy confirmation is PAG, a service which is currently deployed with manual sampling and laboratory analysis (IDEXX, 2020) after about 28 days from insemination.

As well as improving the detection of oestrus, and other reproductive disorders, there is also increasing evidence that progesterone data could be of use in genetics. Growing research suggests that certain genetic traits related to fertility and reproduction can be defined using progesterone profiles. Within the dairy industry, breeding cows with favourable genetics is one of the key ways to create a more productive herd and this will benefit from more objective, numerical data of cow status.

Combining activity and progesterone analysis

As progesterone sensors are single use, there is a finite cost to every sample analysed, whereas a collar or tag has no marginal cost to capture new daily data. Since most cows, particularly with robots, wear an ID system which can cheaply incorporate behavioural monitoring, it makes sense to have an integrated strategy. This might operate by creating a state model for the cow with different sampling strategy for each state and updating the state from data as it arrives.

As there is cow to cow variation in the length of oestrous cycle, a profile can be built up for each cow to adjust these nominal timings. Digital management of cows can also start using a definition of the oestrous cycle based on hours or even minutes since the last event. At present, cows are inseminated in a compromised timing that tries to integrate farm routines and cow availability, perhaps with more precise information we can manage to time inseminations for optimal conception rather than human convenience.

By using hormonal analysis post-insemination, we can remove the costly, disruptive and invasive rectal examinations which are the current fallible method of pregnancy diagnosis.

State	Period days since calving	Sampling method	Purpose
Post-partum	0–20	Progesterone	Detect first ovulation
Cycling	Daily	Behaviour collar or tag	Detect oestrus and map individual cow time of ovulation to end of di-oestrus as this can vary
Follicular	ovulation + 15	Progesterone	Detect change from follicular to luteal predict ovulations ±3 days
Luteal	Alternate days	Progesterone	Confirm luteal phase predict optimal insemination time
Inseminated	Inseminated + 19 days	Progesterone	Confirm conception/pregnancy
Pregnant	Inseminated + 42 days	Progesterone	Confirm conception/pregnancy

State	Period days since calving	Sampling method	Purpose
Pregnant	Insemination + 30 days	PA glycoproteins	Confirm pregnancy

An integrated regime of hormonal and behavioural data to reduce sensing cost and increase precision of timing of insemination

The concept of using combined measurements is an attractive one, since intuitively one might expect results to improve as the number of variables that are included increases. However, there are difficulties with this approach. Firstly, there is the increased complexity of the system particularly if it involves adding extra sensors. Secondly, there is the difficulty of merging the various sets of data. This effectively requires an appropriate weighting to be given to the data from each source. For example, one set of data, say pedometer readings, might, if taken alone, indicate oestrus, whereas another set of data, say milk temperature, does not indicate oestrus. The combined system would be required to attach relative levels of confidence to the two indications to produce a decision. There are established techniques available for data fusion, but whether any of them would be suitable for this application has yet to be investigated. The final main difficulty with the approach is the possibility that some cows at ovulation may not be detectable because they do not exhibit any of the measurable signs.

Discussion: towards a system to improve fertility management

This chapter has attempted to describe the various engineering approaches to the complex biological and managerial problem of identifying the cow fertility status enabling the farmer to get dairy cows pregnant in a timely manner. This long-standing problem has been worked on over decades in all developed dairy economies.

As we improve our understanding of the impact of industrial farming on animal welfare and take into account the changing ethical standards of the consumers, we can see that mutilations of animals and other interventions will become unacceptable to society (Nordquist et al., 2017) and so we can assume that ear tags will become illegal in the future. This review assumed that there could only be a limited role for devices that need to be inserted into the cow either surgically or into the vagina; these devices were reviewed by Senger (1994). Similarly, although synchronisation with injected hormones or inserted devices is widespread and accepted in some territories, it has proved to be an expensive process that goes against the run of consumer preferences for 'natural' methods. This review has looked for devices or systems which can operate with minimal human intervention from outside the cow and can regularly monitor the cow automatically to allow artificial insemination to be carried out at a time when it is most likely to be successful in the natural oestrus cycle. In future, there will need to be systems to identify animals without mutilations which will limit us to as yet unproven video identification and collars and boluses plus the inline monitoring which is very new to the market in 2020.

The existing behavioural technologies have failed to fully solve the fertility problem because they are only detecting a signal of ovulation rather than the event itself and nor can they detect pregnancy. An ideal system for fertility management would be non-invasive to the cow, fully automated, retrofittable to existing milking systems by providing the farmer or fertility technician with predictions of events with close to 100% specificity and sensitivity. In 2020, it appears that inline hormonal analysis may meet this specification and that it can further be enhanced by adding the behavioural data acquired from collars or boluses.

Appendix 1 dictionary of terms in fertility of dairy cows

Anoestrus − cows which show no oestrus activity, either because they are not cycling (true anoestrus) or because oestrus has not been detected (sub-oestrus).

Antepartum − the period prior to birth.

CL (Corpus luteum) − a temporary endocrine structure in the ovary, releases hormones such as progesterone, oestradiol, inhibin A and oestrogen.

DF (Dominant follicle) − also known as **Graafian follicle**, the biggest follicle in ovary and has ability to release a mature egg at ovulation.

Follicular phase − second phase of oestrous cycle, begins after luteolysis, marked by selection of dominant follicle and oestrus, ends with ovulation, controlled by oestrogen.

Follicular waves − the growth of several follicles followed by selection of dominant follicle, occurs two or three times per oestrous cycle, regulated by FSH.

FSH (follicle stimulating hormone) − secreted by anterior pituitary gland, works with LH and initiates follicular growth.

GnRH (gonadotrophin-releasing hormone) − secreted by hypothalamus and is responsible for the release of FSH and LH.

Heat − see oestrus below.

IgG (Immunoglobulin G) − An antibody, and an important component of colostrum.

LH (Luteinising hormone) − secreted by anterior pituitary gland, works with FSH, surge of LH triggers ovulation and development of CL.

Luteal phase − first stage of oestrous cycle, begins after ovulation, includes development of CL (produces progesterone and causes rise in conc.) and ends with luteolysis (triggered by prostaglandin F2α).

Luteolysis − degradation of the CL, marks the end of the luteal phase.

Oestradiol − a type of oestrogen; the most biologically active in oestrous cycle is oestradiol-17 β.

Oestrogen — group of steroid hormones, produced in ovary by follicle, responsible for induction of oestrus.

Oestrous (Am. estrous) — (noun) the main reproductive cycle in female, non-primate mammals (including cows).

Oestrus (Am. estrus) — (adjective) behavioural changes during the pre-ovulation phase of the oestrous cycle, also known as **heat**.

Ovulation — release of egg (ova) from ovary.

Postpartum — the period following birth.

Progesterone — steroid hormone, released by corpus luteum, stimulates uterus to prepare for pregnancy.

$PGF_{2\alpha}$ (prostaglandin F2α) — terminates the luteal phase, resulting lysis of CL and the return to oestrus.

Steric Hindrance — steric effects are non-bonding interactions that influence the shape and reactivity of ions and molecules. Steric effects complement electronic effects, which usually dictate shape and reactivity. Steric effects result from repulsive forces between overlapping electron clouds. As the progesterone particle is tiny in comparison to the lipid bundles in milk, it is hard to ensure that they can find the binding sites on biosensors. Wikipedia.

Appendix 2 Definitions of Accuracy

The results of evaluations of devices are expressed in various ways. Terms such as success rate and accuracy are sometimes used, without definitions. The most useful evaluations are probably those which refer to well-defined quantities such as follows:

Sensitivity $= TP/(TP + FN)\%$

Selectivity or Specificity $= TN/(TN + FP)\%$

False rate $= FP/(TP + FP)\%$

where, TP $=$ the number of true positives.

TN $=$ the number of true negatives.

FP $=$ the number of false positives.

FN $=$ the number of false negatives.

Sensitivity is the proportion of actual positives that are detected by the system; specificity and false rate are measures of the ability of the systems to reject false indications. In general, increasing the sensitivity of a system decreases selectivity, and conversely decreasing sensitivity increases selectivity. For example, a system which generates a large number of false positives will have a low selectivity, but it will probably have a high sensitivity (since there will be few false negatives).

Acknowledgements

A special thanks to Professor Claire Wathes at the RVC, London, for her help with the reproductive biology, Francesca Harding who brought the references in an earlier draft up to date and thanks to Dr. Malcolm Mitchell who told me about the Greek linguistic origins of oestrus.

References

Agscent. (2020). https://agscent.com/team.

Ball, P. J., & Jackson, P. S. (1979). The fertility of dairy cows inseminated on the basis of milk progesterone measurements. *British Veterinary Journal, 135*, 537—540.

Blanchard, T., Kenney, D., Garcia, M., Kristula, M., Wolfer, J., & Haenlein, G. (1987). Relationship of declines in grain consumption and milk-yield to oestrus in dairy-cattle. *Theriogenology, 28*, 407—415.

Blazquez, N. B., French, J. M., Long, S. E., & Perry, G. C. (1988). A pheromonal function for the perineal skin glands in the cow. *Veterinary Record, 123*, 49—50.

Blom, J. Y., & Ridder, C. (2010). *Reproductive management and performance can be improved by use of de Laval Herd Navigator.* Toronto, Canada: First North American Conference on Precision Dairy Management, 2010.

Britt, J. H., Scott, R. G., Armstrong, J. D., & Whitacre, M. D. (1986). Determinants of estrous behavior in lactating Holstein cows. *Journal of Dairy Science, 69*, 2195—2202. Department of Animal Science Box 7621 North Carolina State University Raleigh 27695-7621.

Bruyère, P., Hétreau, T., Ponsart, C., Gatien, J., Buff, S., Disenhaus, C., Giroud, O., & Guérin, P. (2012). Can video cameras replace visual estrus detection in dairy cows? *Theriogenology, 77*(3), 525—530. ISSN 0093-691X.

Carlsen, T. N., Christiansen, P., Gudmundsson, K. F., & Sonnenborg, F. N. S. (2004). *Apparatus for analysing fluid taken from a body patent number: 826299 filed: May 19, 2004.*

Claycomb, R. W., & Delwiche, M. J. (1998). Biosensor for on-line measurement of bovine progesterone during milking. *Biosensors and Bioelectronics, 13*(11), 1173—1180.

Eriksson, S., Jonas, E., Rydhmer, L., & Röcklinsberg, H. (2018). Invited review: Breeding and ethical perspectives on genetically modified and genome edited cattle. *Journal of Dairy Science, 101*, 1—17. Elsevier.

Farmers weekly. (2017). https://www.fwi.co.uk/livestock/livestock-breeding/fertility-production-dairy-farms-latest-nmr-study-shows.

Foote, R. H. (2010). The history of artificial insemination: Selected notes and notables-. *Journal of Animation Science, 80*, 1—10.

Forde, N., Beltman, M. E., Lonergan, P., Diskin, M., Roche, J. F., & Crowe, M. A. (2010). Oestrous cycles in *Bos taurus* cattle. *Animal Reproduction Science, 124*, 163—169.

Foulkes, J. A., Cookson, A. D., & Sauer, M. J. (1982). Artificial insemination of cattle based on daily enzyme immunoassay of progesterone in whole milk. *Veterinary Record, 111*, 302—303.

Foulkes, J. A., & Goodey, R. G. (1988). Fertility of Friesian cows after insemination on the second, third and fourth days of low milk progesterone concentrations. *Veterinary Record, 122*, 135.

Fricke, P. M., & Lamb, G. C. (2002). Practical applications of ultrasound for reproductive management of beef and dairy cattle. In *Proceedings, the applied reproductive strategies in beef cattle workshop, September 5—6, 2002, Manhattan, Kansas.*

Giordano, J. O., Fricke, P. M., Wiltbank, M. C., & Cabrera, V. E. (2011). An economic decision-making support system for selection of reproductive management programs on dairy farms. *Journal of Dairy Science, 94*(12), 6216—6232. https://doi.org/10.3168/jds.2011-4376

Gordon, I. (1996). *Controlled reproduction in cattle and buffaloes.* Wallingford: Publishing CAB International.

Gwazdauskas, F. C., Nebel, R. L., Sprecher, D. J., Whittier, W. D., & Mcgilliard, M. L. (1990). Effectiveness of rump-mounted devices and androgenized females for detection of oestrus in dairy-cattle. *Journal of Dairy Science, 73*, 2695−2970.

Hurnik, J. F., Webster, A. B., & DeBoer, S. (1985). An investigation of skin temperature differentials in relation to oestrus in dairy cattle using a thermal infrared scanning technique. *Journal of Animal Science, 61*, 1095−1102.

IDEXX. (2020). *Pregnancy testing with PAG.* https://www.youtube.com/watch?v=bQ2kcN8LoQU.

Kamphuis, C., Frank, E., Burke, J. K., Verkerk, G. A., & Jago, J. G. (2013). Applying additive logistic regression to data derived from sensors monitoring behavioral and physiological characteristics of dairy cows to detect lameness. *Journal of Dairy Science, 96*(11), 7043−7053. https://doi.org/10.3168/jds.2013-6993. ISSN 0022-0302.

Kiddy, C. A., Mitchell, D. S., & Hawk, H. W. (1984). oestrus-related odors in body fluids of dairy cows. *Journal of Dairy Science, 67*, 388−391.

Klemm, W. R., Hawkins, G. N., & De Los Santos, E. (1987). Identification of compounds in bovine cervico-vaginal mucus extracts that evoke male sexual behavior. *Chemical Senses, 12*, 77−87.

Klemm, W. R., Rivard, G. F., & Clement, B. A. (1994). Blood acetaldehyde fluctuates markedly during bovine oestrous cycle. *Animal Reproduction Science, 35*, 9−26.

Koeslch, R. K., Aneshansley, D. J., & Butler, W. R. (1994). Milk progesterone sensor for application with dairy cattle. *Journal of Agricultural Engineering Research, 58*, 115−120.

Lewis, G. S., & Newman, S. K. (1984). Changes throughout oestrus cycle of variables that might indicate oestrus in dairy-cows. *Journal of Dairy Science, 67*, 146−152.

Ma, W., Clement, B. A., & Klemm, W. R. (1995). Cyclic changes in volatile constituents of bovine vaginal secretions. *Journal of Chemical Ecology, 21*(12), 1895−1906. https://doi.org/10.1007/BF02033850

Maatje, K., De Mol, R. M., & Rossing, W. (1997). Cow status monitoring (health and oestrus) using detection sensors. *Computers and Electronics in Agriculture, 16*, 245−254.

Maatje, K., Rossing, W., & Wiersma, F. (1987). Temperature and activity measurements for oestrus and sickness detection in dairy cattle. In *Proceeding of 3rd symposium automation in dairying, Wageningen, The Netherlands, September 9-11, 1987, IMAG, Mansholtlaan 10−12, Wageningen, The Netherlands* (pp. 176−184).

MacArthur, A. J., Easdon, M. P., & Gregson, K. (1992). Milk temperature and detection of oestrus in dairy cattle. *Journal of Agricultural Engineering Research, 51*, 29−46.

Mazeris, F. (2010). *DeLaval herd navigator proactive herd management.* Toronto, Canada: First North American Conference on Precision Dairy Management, 2010.

McFarlane, N. J. B., & Schofield, C. P. (1995). Segmentation and tracking of piglets. *Machine Vision and Applications, 8*, 187−193.

McLeod, B. J., Foulkes, J. A., Williams, M. E., & Weller, R. F. (1991). Predicting the time of ovulation in dairy cows using on-farm progesterone kits. *Animal Production, 52*, 1−9.

McLeod, B. J., & Williams, M. E. (1991). Incidence of ovarian dysfunction in post partum dairy cows and the effectiveness of its clinical diagnosis and treatment. *Veterinary Record, 128*, 121−124.

Mitchell, R. S., Sherlock, R. A., & Smith, L. A. (1996). An investigation into the use of machine learning for determining oestrus in cows. *Computers and Electronics in Agriculture, 15*, 195−213.

Mottram, T. T. (2016). *Milk analyser and method, patent application W2017/144913.* Priority date 26 February 2016.

Mottram, T. T., Hart, J., & Pemberton, R. (12–16 February 2000). A sensor based automatic ovulation prediction system for dairy cows. In *Proceedings of the 5th national conference on sensors and microsystems Lecce, Italy.* Singapore: World Scientific Publications of Singapore.

Mottram, T. T., Lark, R. M., Lane, A., Wathes, D. C., Persaud, K. C., Swan, M., & Cooper, J. M. (2000). In J. W. Gardner, & K. C. Persaud (Eds.), *Techniques to allow the detection of oestrus in dairy cows with an electronic nose. Electronic noses and olfaction: Proceedings of the 7th international symposium on olfaction and electronic noses.* Brighton, UK: Institute of Physics, 2000.

Nebel, R. L. (1988). On-farm milk progesterone tests. *Journal of Dairy Science, 71,* 1682–1690.

Nordquist, R. E., van der Staay, F. J., van Eerdenburg, F. J. C. M., Velkers, F. C., Fijn, L., & Arndt, S. S. (2017). Mutilating procedures, management practices, and housing conditions that may affect the welfare of farm animals. *Implications for Welfare Research, Animals (Basel), 7*(2), 12. https://doi.org/10.3390/ani7020012. Published online 2017 Feb 21.

Norman, H. D., Wright, J. R., Hubbard, S. M., Miller, R. H., & Hutchison, J. L. (2009). Genetic and environmental factors that affect gestation length in dairy cattle. *Journal of Dairy Science, 92,* 2259–2269. https://doi.org/10.3168/jds.2008-1768

O'Sullivan, M., Shalloo, L., Pierce, K. M., & Buckley, F. (2020). Economic assessment of Holstein-Friesian dairy cows of divergent Economic Breeding Index evaluated under seasonal calving pasture-based management. *Journal of Dairy Science, 103*(11), 10311–10320. https://doi.org/10.3168/jds.2019-17544

Paisley, L. G., Mickelsen, W. D., & Frost, O. L. (1978). A survey of the incidence of prenatal mortality in cattle following pregnancy diagnosis by rectal palpation. *Theriogenology, 9,* 481–491, 1978 - Elsevier.

Pederson, S., & Pedersen, C. B. (1995). Animal activity measured by infrared detectors. *Journal of Agricultural Engineering Research, 61,* 239–246.

PennState Extension. (2016a). *Video, characteristics of the bovine oestrous cycle.*

PennState Extension. (2016b). Milk progesterone analysis for determining reproductive status.

Peter, A. T., & Bosu, W. T. K. (1986). Postpartum ovarian activity in dairy-cows – correlation between behavioural oestrus, pedometer measurements and ovulations. *Theriogenology, 26,* 111–115.

Peters, A. R., & Ball, P. J. H. (1994). The ovarian cycle, oestrous behavior and its detection. In *Reproduction in cattle* (pp. 23–62). Oxford: Blackwell Science.

Rajamahendran, R., Robinson, J., Desbottes, S., & Walton, J. S. (1989). Temporal relationships among oestrus, body temperature, milk yield, progesterone and luteinizing hormone levels and ovulation in dairy cows. *Theriogenology, 31,* 1173–1182.

Rasmussen, D. C., & Carlsen, T. N. (2017). *A tape for biomarker analysis of a milk sample, Patent Application WO2019/132761.*

Schofield, S. A. (1988). *Oestrus in dairy cows.* Public University College of North Wales.

Schofield, S. A., Phillips, C. J. C., & Owens, A. R. (1991). Variation in the milk-production, activity rate and electrical-impedance of cervical-mucus over the oestrus period of dairy cows. *Animal Reproduction Science, 24,* 231–248.

Senger, P. L. (1994). The oestrus detection problem: New concepts, technologies and possibilities. *Journal of Dairy Science, 77,* 2745–2753.

Shahriar, M. S., Smith, D., Rahman, A., Freeman, M., Hills, J., Rawnsley, R., Henry, D., & Bishop-Hurley, G. (2016). Detecting heat events in dairy cows using accelerometers and unsupervised learning. *Computers and Electronics in Agriculture, 128,* 20–26. ISSN 0168-1699.

Stevenson, J. S. (July 2016). Synchronization and artificial insemination strategies in dairy herds. *Veterinary Clinics of North America: Food Animal Practice, 32*(2), 349−364. https://doi.org/10.1016/j.cvfa.2016.01.007. Epub 2016 Apr 1.

Stevenson, J. S., Smith, M. W., Jaeger, J. R., Corah, L. R., & Lefever, D. G. (1996). Detection of oestrus by visual observation and radiotelemetry in peripubertal, oestrus-synchronized beef heifers. *Journal of Animal Science, 74*, 729−735.

Tillett, R. D., Onyango, C. M., & Marchant, J. A. (1997). Using model-based image processing to track animal movements. *Computers and Electronics in Agriculture, 17*, 249−261.

Van Vliet, J. H., & Van Eerdenburg, F. (1996). Sexual activities and oestrus detection in lactating Holstein cows. *Applied Animal Behaviour Science, 50*, 57−69.

Velasco-Garcia, M. N., & Mottram, T. T. (2003). Biosensor technology addressing agricultural problems. *Biosystems Engineering, 84*(1), 1−12.

Voronin, V., Brayer, E., & Ben Menachem, U. (2005). *A method and device for detecting estrus, IL166394D0*.

Vreeburg, N. (2010). *Precision Management on two Dutch dairy farms by use of Herd Navigator*. Toronto, Canada: First North American Conference on Precision Dairy Management, 2010.

Warren, M. E. (1984). Biological targets for fertility and their effects on herd economics. In R. J. Eddy, & M. J. Ducker (Eds.), *Dairy cow fertility* (pp. 1−14). London: British Veterinary Association.

Wathes, D. C. (2021). Care of the post-weaning calf 150. In S. Mahendran (Ed.), *Handbook of Calf Health and Management*. Sophie Mahendran and Claire Wathes Pub, ISBN 9781789181340.

Williams, M. E., & Esslemont, R. J. (1993). A decision support system using milk progesterone tests to improve fertility in commercial dairy herds. *Veterinary Record, 132*, 503−506.

Williamson, N. B., Morris, R. S., Blood, D. C., & Cannon, C. M. (1972). A study of oestrus behaviour and oestrus detection methods in a large commercial dairy herd. I. The relative efficiency of methods of oestrus detection. *Veterinary Record, 91*, 50−58.

Further reading

Boyd, H. (1992). Oestrus and oestrous cycles: Problems and failures. In A. H. Andrews, R. W. Blowey, H. Boyd, & E. G. Eddy (Eds.), *Bovine medicine* (pp. 433−448). Oxford: Blackwell Science.

Darwash, A. O. (1996). The importance of progesterone levels during early pregnancy in the cow. In *Nottingham cattle fertility conference 1996*. University of Nottingham.

Mahendran, S. (2021). *Handbook of calf health and management. ISBN 9781789181340 from chapter 7, care of the post-weaning calf 150, Sophie Mahendran and Claire Wathes*. Publishing 5m Books Ltd ©.

Pemberton, R. M., Hart, J. P., & Mottram, T. T. (2001). An electrochemical immunosensor for milk progesterone using a continuous flow system. *Biosensors and Bioelectronics, 16*, 715−723.

Rivard, G. F., & Klemm, W. R. (1992). Gas Chromatographic analysis and oestrous diagnostic potential of headspace sampling above bovine body fluids. In R. L. Doty, & D. Muller-Schwarze (Eds.), *Chemical signals in vertebrates VI* (pp. 115−118). New York: Plenum Press.

Resilient food infrastructure and location-based categorisation of urban farms

Negin Minaei[1,2]

[1]*Urban Studies Program, Innis College, University of Toronto, Toronto, ON, Canada;* [2]*CITY Institute, York University, Toronto, ON, Canada*

Abbreviations

CEA	Controlled Environment Agriculture
CEP	Controlled Environment Production
CHCI	Canada Healthy Communities Initiative
CUI	Canadian Urban Institute
G.A.P	Good Agricultural Practices
GHGs	Greenhouse Gases
GTA	Greater Toronto Area
i-RGT	Integrated Roof Top Greenhouse
UHI	Urban Heat Islands
UN	United Nations
WWII	World War Two

Introduction

Population overgrowth has put cities and the planet under a lot of pressure. Feeding people in cities has become the new concern and has encouraged many planners to think about food growing in cities. They have defined new city concepts such as 'Eco Cities' and 'Regenerative Cities' with a focus on sustainable agriculture. In 2018, approximately 55% of the world's population lived in urban environments; United Nations projected that number would reach to 68% by 2050 (UN, 2018). The share of agricultural energy and water consumption for food production is considerable. It seems that agricultural environmental footprint is farm specific (Adewale et al., 2019) and that is perhaps the reason that most studies refer to the assumption of Kulak which is *%20 to %30 of the global GHG emissions is caused by agriculture industry* (Kulak et al., 2013 In Eigenbrod and Gruda, 2015), whilst EPA (2021) reports a 10% emission by this sector. That has been the major drive for global calls to

Digital Agritechnology. https://doi.org/10.1016/B978-0-12-817634-4.00003-3

action for climate-smart agriculture to reduce the GHG emissions and environmental footprint. With the acceleration of global warming and climate change, Li and Siddique (2018) predicted that agricultural productions would decrease by 2% a decade, whilst food demand would increase by 14%. They consider any types of grains that could be grown in a dry land would be a climate-smart food choice. The other important factor is the environmental footprints which span from water and energy footprint to carbon and nitrogen footprint (Karwacka et al., 2020); in some cases, methane footprint is measured too. Other factors that impact the environment are deforestation rate, pesticide usage and animal welfare, but they are not counted in measuring the footprints. Karwacka et al. provided a review on each of these footprints and showcased the best practices by different countries to minimise them; for example, Spain has been considering to impose taxes on the food items with largest carbon footprints mainly the meat category because livestock has been identified as the most resource-intensive food produce with variety of environmental footprints.

Since the emergence of Sustainable Cities concept and Sustainable Development Goals by the United Nations, food security and food production have gained far more importance than before particularly in urban areas as urban dwellers are the main consumers and wasters of food. At least 6 goals of the 17 Sustainable Development Goals are somehow related to food including Goal 2 (Zero Hunger), Goal 6 (Clean Water and Sanitation), Goal 11 (Sustainable Cities and Communities), Goal 12 (Responsible Consumption and Production), Goal 13 (Climate Action), Goal 14 (Life below Water) and Goal 15 (Life on Land). Goal 2 aimed to ensure that resilient and sustainable food production systems and practices with higher agricultural productivity were available to all people by 2030 to maintain the ecosystem, and help communities adapt with severe disaster, weather events and climate change.

Considering the changes that COVID-19 has brought to cities, mainly changing the mentality of living in a single home and distancing from the downtown cores to have a safer air space, we cannot predict the future percentage of people who choose to live in the suburbs. The pandemic and lockdowns have increased our awareness of the broken food production system in cities; as Pulighe and Lupia (2020) have elaborated nicely, cities have been the consumption sites for water, food, energy and all sorts of resources. Regardless of residence location, most people still do not grow their own food and are dependent on cities to provide their food for them. COVID-19 pandemic revealed many serious issues in the food production systems for cities. Lockdowns and pauses in export and import of foods, interruptions to transportation networks and labour shortage caused serious concerns for governments to ensure that food security and supply—demand equation were still balanced and properly working. It brought to light the fundamental vulnerability of cities and their severe dependence on external sources to provide them the essential and perishable goods such as fresh vegetables and fruits. In addition, it showed us that we need to grow food in the country and not rely on importing food from other countries as our survival depends on food. The attempt of UK supermarkets to calm down the fear of food shortage after disruptions to the freight shipping and border chaos during December 2020 is another example that attracts our attention to the importance of producing food locally rather than relying to import it internationally (Thomas, 2020). In Canada, early in spring, we learned that most of the farm workers were temporary migrant workers who used to come from Mexico and some South American countries and due to the COVID-19 restrictions they were banned from travelling which made Canadian farm owners extremely worried in a way that they believed in fall 2020 Canada would experience serious food shortage. Worries of not having access to enough food to feed the population during the fall and winter times and before each lockdown were at their highest level during the spring 2020 and caused extreme reactions by people such as panic buying behaviours (Hobbs, 2020) of some particular food products and at the same time large-scale waste of fresh produce by farmers due to the closure of restaurants and bars which were their main customers. Also small growers could not sell their produce to their local restaurants.

A serious problem is that available agricultural lands suitable for farming are lost by urban expansion and development. For example in the GTA, Ontario, in the Greater Golden Horseshoe such as the Oakridge region, all forest and farm lands are currently being sold to develop real estate. The speed of losing agricultural lands in urban areas is accelerating. Many fruits and vegetables are imported from the Far East and West, whilst they could be grown here in Ontario (CUI, 2019). According to the Ontario Farmland Trust, urban expansion in the Greater Toronto Area led to the loss of 2000 farms and about 60,700 ha (150,000 acres) of farmland between 1976 and 1996 and another 20,000 ha (50,000 acres) between 1996 and 2006 and between 2011 and 2016 about 2.5% drop in farmlands (CUI, 2019).

Solutions such as vertical farming and aquaculture have spread all around the globe. Sustainable city practices have also introduced food growing in cities by allowing communities to use land owned by others such as community gardens in parks and city-owned lands and growing food on roof tops. Researchers recommend that we shall benefit from all sorts of available technologies to ensure cities have edible green infrastructure by integrating vertical farming and more soil and water conservation solutions such as hydroponics, aeroponics and aquaponics (Pulighe & Lupia, 2020).

Urban agriculture and categorisation of structures

Cities confront many challenges and one of the most dangerous ones that can harm people is the Urban Heat Island (UHI) effect. Since studies on cities' climates proved that 'Eco Roofs' such as 'White Roofs' and 'Green Roofs' can decrease the temperature dramatically, green roofs and green walls have found a special place in the design of sustainable cities and urban environments. Literally, these green surfaces have been opted as a sustainable solution to decrease the inner city temperature and the negative effects of UHIs. Many cities have started greening their exterior walls with beautiful green plants. European Confederation of Green Roofs and Walls (n.d.) reports these benefits for green surfaces:

- improving urban micro-climate
- improving the air quality and decreasing air pollution
- acting as a thermal insulation
- acting as sound insulation and decreasing noise pollution
- retaining water and increasing humidity
- improving biodiversity
- acting as a shield and protecting facade material against UV radiation, chemical pollutants and dirt and therefore prolonging the buildings' life-cycle
- providing a nicer view and therefore adding value to the building

This practice has been common in most cities regardless of their attention to the sustainability, resilience and ways to confront climate change and UHI. In some cities, it has come as an architectural trend to green facades and the roof of buildings. But after a whilst, new directions found their ways into greening cities which is generally called 'urban agriculture'. Geographers define it as any form of activity to produce food within or close to cities, whilst different groups of people and stakeholders with various reasons, motivations and socio-economic conditions are involved (Duží et al., 2017). For example, instead of having a merely green roof covered with vegetation or a hybrid roof of vegetation and solar panels, people used the roof tops in their residential towers as community gardens to produce food. Paying attention to the motives and possible profits, three types of urban food gardening have been categorised, a small scale non-for-profit gardening, urban farming and non-urban oriented farming (Simon-Rojo et al. 2015 In Duží et al., 2017). In this chapter, our focus is at the context and place of farms in cities, and the types of technologies they use to produce more food accessible to city dwellers.

Growing food for large populations in cities requires land, energy and water which are often not available. Most cities including the City of Toronto and particularly Live Green Toronto educate the public to grow their own foods. With different incentives, they encourage their citizens to use natural light and rain water to plant herbs and vegetables in their gardens, and even in their balconies. PollinateTO (an initiative from the City of Toronto) aims to improve the pollinators' population, so they raise awareness on pollinator gardens and teach the public to shrink their lawns and use native drought-resilient ground covers instead. They share with them the best practices of growing pollinator gardens and rain gardens or using boulevard lawns to grow food both individually and with their neighbours in their communities. PollinateTO grant incentive is a good example of the types of actions being implemented to fulfil the pollinator protection strategy (City of Toronto, 2019).

One solution is to employ Controlled Environment Production (CEP) and farming within inner city areas to ensure less GHG emission is produced by decreasing energy and water consumption, less prepping, packaging and shorter distances to travel. One of the best examples of controlled-ecosystem is the InFarm modular system which received the Global G.A.P certification as the first hydroponic farming system that developed a standard for sustainable and safe food production. Their system is now used in different countries such as Switzerland,

Denmark, France and Germany (InFarm, 2020). They reduced food transport by 90%, water consumption by 95% and pesticide by 100%, all because they could set up their modular systems in shopping centres, schools, hospitals and restaurants.

With the advancement of information technologies and telecommunication, people are sharing their experiences of food growing and therefore they are more aware of the best practices of urban agriculture. There have been many trends in food growing in urban areas that have considered mass food production and in some cases even sustainable and self-sufficient practices. Here by self-sufficient I mean collecting rainwater, getting natural sunshine and therefore not consuming electricity for heating or irrigation. All can be done naturally in a semi-controlled indoor environment. Please note, I consider dependence on active technologies and power, which may be shut down at once, as a weakness because they do not contribute to resilience of a system.

I have categorised different types of urban farms in Table 8.1 and briefly describe their positive and negative characteristics.

Growing over the ground

Green roofs or rooftop gardens

Green roofs have been identified as an effective solution to the UHI phenomena to help cities cool down in the inner city areas. In the early forms, they were only a layer of light vegetation covering the roof, but in recent years they have transformed to community gardens and rooftop gardens, a common place for residents of the building to come together and grow their own food. Green roofs can help with absorption of heavy rains and growing a micro-ecosystem that can encourage biodiversity. With the right type of plants and flowers, they improve the quantity and quality of pollinators which are essential to grow fruits and vegetables.

Perhaps the LIFE@UrbanRoofs is a good example of benefiting from rooftops to create outdoor spaces as well as roof gardens. It was an initiative to answer to the Horizon 2020 as a program that could bring resilience and sustainability to the future smart cities. The famous Peperklip building in Rotterdam with 7600 square meters green roof — as the longest green roof ever built — is one of the three locations that were assigned to roof gardens. The LIFE project received a subsidy from the European Commission to support nature and environmental conservation as well as climate action across Europe (Rotterdam, 2019).

Other notable examples are the airport rooftops that converted to urban farms. The two famous examples are the JetBlue's T5 farm which is located at the departures level of the airport with roughly 24,000 square foot of rooftop and the Eero Saarinen's 1960 TWA terminal building at the New York's John F Kennedy airport (Burnham, 2020a). Producing fresh local vegetables at the airport means affordable food prices available at the restaurants and cafes of the airport.

Table 8.1 Categorisation of urban farms and their characteristics.

	Dependence on power	Self-sufficient	Horizontal	Vertical	Costly set up
Overground farms					
Green roofs/rooftop gardens		✔	✔	✔	
Community gardens		✔	✔	✔	
Brown-fields, derelict lands and buildings		✔	✔	✔	✔
Green towers and buildings	✔		✔	✔	✔
Public places such as schools and airport		✔	✔	✔	
Underground farms					
Walipini		✔	✔	✔	
Sunken greenhouse		✔	✔	✔	
Underground tunnels and shelters	✔		✔	✔	✔
Tunnels through mountains	✔		✔	✔	✔

Source: the author.

Community gardens

Since the main reason for the existence of community gardens in cities is producing fresh food in neighbourhoods and local areas, urban allotments have found favour amongst those who do not have a garden of their own. They have found attention from city officials and politicians because it is considered a sustainable development practice, and not only it brings food to the cities, it improves the social environmental sustainability of a society. Some predicted that it would be incorporated into urban planning and urban design soon (Duží et al., 2017). Productive landscapes in cities are not a new topic; they are essential part of the green infrastructure cities required to stay self-sustained and resilient. Some researcher like Siegner et al. (2018) question the positive contribution of urban-produced food to food access and food security of low-income communities in US cities. Their rational is that there have not been proper studies to measure the real impact of these small practices and most of the literature is theoretical considering the potentials of urban agriculture. Another point is that small space cannot produce enough food access and bigger spaces are often privately owned and do not serve the lower index of the community. If we add the matters of connecting to consumers and logistics, it gets even more complicated and costly for small commercial growers. Their critics can be right, as a considerable portion of the urban-produced food examples that I read belonged to the richer investors who produced a large amount of high-quality food. Often their farms are equipped with technologies and their customers are chefs and top-notch restaurants. However, Goodman and Minner (2019) identified that institutional Controlled Environment Agriculture (CEA) can relatively benefit the job market, low-income communities and health and well-being of food-insecure families. They call the CEA a form of urban farming that is globally on the rise in cities.

There are different types of community gardens; if the land is owned by the city, growers often do not pay any rent. If the land is owned privately, growers pay some rent. Fig. 8.1 shows a community garden where the public can rent a few plots of land for gardening.

The new practice in the United States, which was in the spotlight whilst ago, was about a young man called Rob Greenfield who lived (rent-free) in the back garden of a resident and grew food in the neighbour's front gardens in that neighbourhood and gave them their share. He was experimenting extreme sustainability and his project was food freedom. He as an entrepreneur could secure himself a job (keeping front gardens of their neighbours clean) and grow fresh vegetables and herbs in those pieces of lands, again without owning anything (Reynolds, 2019).

FIGURE 8.1

Community garden by George Tan, CC0 1.

Source: https://search.creativecommons.org/photos/a96608ac-e4ec-429b-a52d-810227383eee.

In the recent $31 million dollar Canadian government initiative which is called 'Canada Healthy Communities Initiative', one acceptable project category is the community garden (CHCI, 2021). After the pandemic, food banks have been in trouble and called for help. Food poverty has become a serious issue as many people have lost their jobs and their basic incomes. Community gardens can bring people together and lead to social sustainability. They can help them fight against food poverty. Although federal governments are keen supporters of such initiatives, in reality when it comes to municipal governments, they are not that collaborative despite their inspirational pronouncements. I have been involved with a community group to establish a community garden in our district and I know that they have been communicating with the Parks Department of the City of Toronto to get a vacant piece of land that has been derelict for years; after over 2 years and a half, they were offered another parcel and are supposed to work under the supervision of an assigned person, still in communication stage. The City of Toronto is definitely an active advocate of food growing in Toronto and I as a volunteer have worked with the Live Green Toronto under the Environment and Energy Department of the City since 2015. I was trained to raise environmental awareness on food growing and in all of our outreach events, we taught the public how to do that. In reality, it is not as easy as it sounds.

Brownfields, derelict lands and vacant buildings

Community-based gardens in derelict lands in urban areas where people gather together to grow food often have similar type of problems. Vandalism and theft of the produce is often mentioned by food growers and since the land does not have any kind of walls or borders and it is an open public urban space, most growers often plan to have some kind of presence on site or to personalise their garden beds to show they belong to them.

When a company or a person takes a derelict piece of land and turn it to a commercial farm, everything changes including its security. For example, GrowUp Urban Farms turned previously unused brownfield sites into productive areas of aquaponics to farm fish and grow leafy greens in a soil-less system. Another example is Farmizen which is a company in Hyderabad, India, which rents 55 square meters mini-farm to city dwellers to grow their own organic produce. They have set up a mobile-based platform which is currently live and allows users to go and visit the farm any time they want. Their developed app has 1500 subscribers and 161,874 square meters of land under cultivation (Oyuela, 2019) (Fig. 8.2).

Abandoned lands under the highways and bridges are often places where homeless people set up their tents at least in Toronto before they are securely removed from the area by the City of Toronto's Police. In other cities, people use those parcels of land to set up their rent-free community gardens. Fig. 8.3 shows one.

Using derelict buildings such as warehouses and industrial spaces to create smart farms in which plants could be protected from pollutants, pest and bad weather is another example of using indoor gardens. A famous example is Plenty, a vertical farm in a warehouse in South San Francisco which has been invested in by Jeff

FIGURE 8.2

Amersham Road Community Garden (1), Photo by Karen Blakeman is marked with CC0 1.

Source: https://search.creativecommons.org/photos/4eeb2050-db35-46a3-b6db-4b5f57cf0474.

FIGURE 8.3

Abandoned space under Vancouver's Skytrain which is now used for gommunity garden by Roland Tanglao, marked with CC0 1.0.

Source: https://search.creativecommons.org/photos/81618780-982c-4358-bdc0-14473911506e.

Bezos and Google's chairman, Eric Schmidt (Venture City, 2019). A famous American example is the vertical farm called Bowery Farming that stacked crops vertically to benefit from the maximum space to grow inside an industrial space which is equipped with modern technologies such as real-time monitoring of quality and safety of plants and nurturing them (Bowery, 2020). Their vision is to grow this system of gardening in cities to ensure city dwellers have access to fresh local herbs and leafy greens year round. They select industrial spaces outside of cities and turn them into smart farms. They have extended their community in Nottingham, Maryland, and Baltimore and have contributed to food banks based on their mission of fighting food insecurity in cities.

Another example of using derelict buildings in cities is the Gotham Greens which used the former steel mill called Bethlehem Steel Plant and turned it into a 9290 square meters hydroponic greenhouse which is the first commercial rooftop greenhouse in the United States. It uses 100% renewable energies and recycled irrigation water (Bozhinova, n.d.). Most of these types of start-ups partner with whole food markets, local organisations, food secure organisations and non-profits and donate seedlings to local programs.

Vertical growing and green walls

Due to the lack of arable lands in cities and the growing demand for fresh produce and the importance of food security in urban areas, different kinds of design for vertical growing have emerged for indoor and outdoor surfaces. First of all, it is important to ensure we know the differences between all forms of vertical growing including the following: vertical gardens versus vertical farms and living walls (plant walls) versus green facades. Alter (2018) has described and illustrated their difference nicely as their systems and products are different. Our focus in this chapter is on food growing, but it's important to know there are examples of growing food on facades and living walls too. Bright Agrotech is one of them, a US-based company that has developed a hydroponic farm system in the form of simple lightweight panels that are self-watering, eliminating the weeds and can attach to any wall anywhere in all urban spaces, along the sidewalks or in the back alleys. Because of its characteristics, it consumes less time and energy to produce green leaves (Peters, 2015). In most populous countries across the globe, vertical gardening with the most advanced technologies and often hydroponic systems has become the common solution to feed city inhabitants; for example, in Singapore's downtown, the Sustenir Agriculture provides almost a ton of kale and 3.2 tons of lettuce per month employing smart technologies.

Kalantari et al. (2017) emphasise on the importance of vertical farming technologies as they see how architectural technology and farming technology should perfectly work together to achieve an optimal space for food growing. In their paper, they looked at different international case studies and reviewed technologies and even provided an implementation guide which discusses structure, material, lighting, natural lighting, solar cells, water requirements and recycling systems, and methods of replacing electricity with solar energy. They discuss how livestock production is possible in vertical farming too. Beacham et al. (2019) reviewed vertical farming categories and divided them to two categories: first, those that are grown on multiple horizontal growing platforms and second, those that climb up on vertical surfaces. They also called all rooftop greenhouses and single-level platforms assigned to food growing in cities the 'Building Integrated Agriculture' and did not separate them apart from the vertical farming category. Their category of vertical farming includes from group 1) stacked horizontal systems, multi-floor towers and balconies and from group 2) green walls and cylindrical growth units. European Confederation of Green Roofs and Walls (n.d.) identifies three types of green walls: Ground-based, plant-trough based and wall-bound. In the first group, plant roots are in the ground and they are either self-climbing plants going up on walls or they climb a trellis. The second group is self-explanatory and the third group consists of wall-based green walls with plant-trough modules or wall-bound systems which are also called living walls.

My categorisation is based on the location of the growth medium, whether it is indoor in a controlled environment or it is outdoor or perhaps installed on a building facade or in urban parks.

Indoor vertical growing

Indoor vertical growing spans from very advanced smart and robotic technologies to less advanced growing in enclosed spaces from buildings to containers. Aerofarms is an example of those smart advanced systems. Using 95% less water and energy and zero pesticides, plus almost 400 times higher productivity compared to traditional farming, they have become a leader in commercial fully controlled indoor vertical farming (Oyuela, 2019). Their smart aeroponics system mists the roots of plants with water, oxygen and nutrients. Like all other smart systems, they have engineered lighting systems mainly made of LEDs to make it energy efficient and which can be tuned to plants favoured wavelengths and their farms benefit from precision farming, machine vision and learning, IoT and data science. Their unique qualities could be the smart pest management, smart scaling and the smart reusable growing substrate that they created which can be sanitised after each harvest (Aerofarm, 2021).

One of the interesting examples of utilising unused space in cities was a program called farm to table office program in Pasona Urban Farm in Tokyo, Japan. Pasona refurbished a 50 years old building, and in addition to office areas, auditorium and cafeterias, it included a rooftop garden and a considerable indoor green space in which 200 species of vegetables, fruits and rice were grown. The double-skin green facade of the building was covered with orange trees and variety of flowers all on its small balconies. Almost in all spaces, different vegetables were growing, tomatoes above conference tables, salad leaves inside seminar rooms, bean sprouts under benches and passion fruits and lemon trees acting as partitions. The food that was grown by the employees was harvested and served in the cafeteria to them. Climatic factors were all monitored to ensure the space was safe for employees (Andrews, 2013).

Fig. 8.4 shows an automated growing farm which was exhibited in the Royal Agricultural Winter Fair in 2015 in Toronto. This vertical unit can be easily installed in any shipping container. It even has a place for pollinators to feed and pollinate the plants inside the unit.

Using unused shipping containers as small urban farms across cities to produce healthy, tasty and affordable food in Paris was an idea of two young entrepreneurs Guillaume Fourdinier and Gonzague Gru who were both the sons of French farmers (Stimmler-Hall, n.d.). They used recycled containers, controlled the environment and equipped it with LED lights mimicking the summer sunshine to encourage faster growth. They could produce 7 tonnes of strawberries annually in 30 square meters whilst saving 90% water. According to the Agricool website, they later expanded their Cooltainers to other cities and countries including Dubai's Sustainable City.

The American version was set up by Kimbal Musk (Elon Musk's brother) who established the Square Roots. He used shipping containers and enabled them with technology and created movable urban farms to grow 'hyper-local', 'real-food' and 'year-round'. Square Roots is based in Brooklyn, New York. The vertical farms inside each container equals 8094 square meters of farm land and they produce 22.67 kilograms of leafy greens per week. Square Roots offers educational services to train future generation of farmers (Square Roots, n.d.). They have set up a plant growing library in which they study the perfect growing conditions for each plant so it can be later used to grow food in space (Venture City, 2019).

Integrated Rooftop greenhouse (iRTG)

Rooftops have been another good alternative for land particularly in cities. Since UHI effect is a serious problem in urban environments and decreasing the city's surface temperature has been identified as an optimal solution, many buildings converted their roofs to roof gardens and community gardens to grow food instead of

FIGURE 8.4

Automated growing farm, the Royal Agricultural Winter Fair 2015, Toronto, Ontario, Canada.

Source: https://www.flickr.com/photos/neginminaei/22428214064/.

merely a cool roof or a green roof covered with some local plants. This approach had shown benefits to residents of those buildings as they can grow their own food. IRTG is defined as an innovative form of vertical farming that combines a greenhouse to a buildings' rooftop whilst benefiting from rainwater collection, residual air and heat and can preserve resources. In a comprehensive study and environmental assessment, Sanjuan-Delmás et al. (2018) have carefully analysed this system and claimed that this can be an optimum option to improve food security in cities compare to conventional production methods. They compared yield and greenhouse emissions and even fertilisers. Their results suggest that IRTG can act in a larger scale if management system is programmed properly and if the infrastructure is optimised to decrease their environmental footprints.

BIGH farms (2018) in Brussels have used the Foodmet market hall rooftop as an outdoor garden to complement their high-tech greenhouse with its aquaponic system. This enabled them to produce fish, fruits, vegetables and herbs in a closed zero waste loop.

Outdoor vertical growing I

The most common form of outdoor vertical growing is green walls also called 'living walls'. After their grand entrance into interior design of common spaces and halls mainly in semi-public spaces, in recent years, they have been used as a medium to set up food growing gardens. In many community gardens, vertical living walls are now producing leafy greens. The fact that in populated cities most people do not have a garden to grow anything and often do not have enough space on their balconies to have pots for food growing, growing on vertical walls has become a reasonable solution to the lack of space problem. Fig. 8.5 shows an example of this kind of green walls.

FIGURE 8.5

'Green Grid' by cogdogblog is marked with CC0 1.0.

Source: https://search.creativecommons.org/photos/d7486cee-8e49-4dde-856c-102e6a67c294.

It has other benefits too. For example, ergonomically, most people are more comfortable to plant, pick or work whilst they are standing rather than bending. This method has responded to a common need so well that in recent years some companies are selling IKEA-like systems components so customers can mantle it themselves. For instance, Varden or Vertical Garden sells living wall systems which has all the needed kits from filled sacks of soil and refills to mounting bars, galvanised wire panels and bottom trays, both for indoor walls or outdoor walls (Varden, n.d.). These types of systems are cost-effective, do not need light to grow or complicated cleaning mechanisms, they have a pump for irrigation and as long as they are installed in a place with direct light, they fulfil their purpose.

Growing Underground

Walipini and Sunken green houses

Walipini in Bolivian language means a place of warmth. Essentially, a Walipini is a greenhouse which is partly under the ground. It was invented by a Swiss volunteer, Peter Iselli, who was funded by a European Development Fund to develop a technology that can stand erratic rainfalls, high UV radiation and increasing temperature for Bolivian farmers who live on the Altiplano (Rosendo, 2018). The easiest form of building it is to dig a cube out of the ground and cover the top with a transparent clear material so it can absorb the Sun's thermal energy and the Earth's geothermal energy. It is particularly very useful for areas with colder climates, as it provides the chance to choose and grow a variety of vegetables all year round (Ashwanden, 2016). It is a passive solar style of a greenhouse benefiting from natural light, natural heating/cooling, natural ventilation and natural irrigation. It can be totally self-reliant without any dependence on electricity; it can rely on rainwater collection. Overall, a very resilient and sustainable option. Ideas of 'underground greenhouse' or 'sunken greenhouse' were inspired by the concept of Walipini. Because it is situated under the ground, it benefits from fewer temperature changes and so has a stable temperature most of the time. It is not costly, because there are no walls to build so the building material costs can be saved (Avis, 2013). Although in the original experiment brick walls were installed, in our case study, some forms of wooden walls are installed. The original Bolivian farmers build four brick walls and plant ivies on them to retain extra moisture.

Figs. 8.6 and 8.7 illustrate the process of building a sustainable and self-sufficient Walipini in case study of Red Gate Farms in the United States. It uses all things natural. Sun provides the lighting and part of the heating. The rain water is collected from the roof and stored in the barrels and later can irrigate the plants (7-3). Under the wooden floor, there is an empty space that acts as a duct. It benefits from the geothermal energy and brings the warm air by a horizontal duct under the ground to the space under the wooden floor so in the winter inside the Walipini is warmer than outside. More information can be found on their website and YouTube Channel. They generously shared their high-quality photos to promote sustainable farming as it has proved to be a successful example.

FIGURE 8.6

The process of building a self-sufficient Walipini by Red Gate Farms.

Benefiting from the geothermal energies in sunken greenhouses means having a stable temperature inside the greenhouse without consuming electricity. Walipini users report about 20 degrees temperature difference between indoor and outdoor which is considerable and can translate to possibility of producing food all year round.

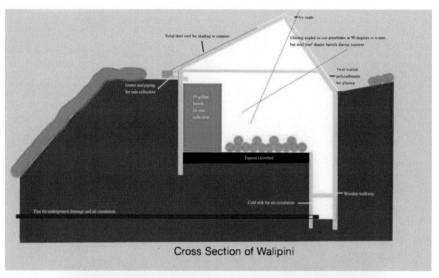

Cross Section of Walipini

FIGURE 8.7

Cross-section and finish of the self-sufficient Walipini by Red Gate Farms.

Source: https://www.redgatefarmllc.com/the_farm/the_walipini.

In South Korea, it's common to use heat pumps to keep the greenhouses warm. Lim et al. (2020) suggest to use air-to-water heat pumps to use underground air to heat the greenhouse. Their analysis showed up to 70% of the total heating costs could be reduced compared to an air heater.

Underground spaces under cities and urban environments

There is no doubt that re-purposing spaces in cities and finding a new function for them is the most sustainable approach that has been opted by architects in recent years. Some call it 'Adaptive Reuse' which mainly applies to buildings which are not used for the purpose they were designed for but they have some kind of value like a historical or industrial heritage value and that prevents their demolition. This approach has found its way to the underground farming as well. Among all benefits of growing food under the ground, these are the highlights: first, the cool and stable temperature of 16°C under the ground provides an ideal consistent environment to grow crops all year round. Second, far below the ground, there is no pest and therefore no pesticide is used. That makes the produce healthier and organically pest-free (Burnham, 2020b).

Plantagon CityFarm is an example of this best practice. It is a farm that was set up in an old newspaper archive underneath a 26 storey office building in Stockholm. They planned to capture the heat from the LED lights installed in their vertical farms and allow it to naturally vent out and heat up the office building using a heat storage system such as underground heat pump system. A pretty clever solution to keep plants cool and keep the office warm conserving energy (Peters, 2017).

Bomb shelters under ground

Perhaps the only air-raid shelter and the most famous of all is the underground farm that was set up in the WWII bomb shelter about 33 m under the Clapham London and was transformed to a Hi-Tech underground farm. Since this is one of the longest tunnels beneath Clapham, it was filled with layers of stacked hydroponic beds forming a vertical farm. The main produces are herbs and salads and it saves both water and energy (Growing Underground, 2015).

Metro farms in subway tunnels

A subterranean subway station in South Korea called Sangdo Station has turned part of its metro station to an organic farm and grows leafy greens, sprouts and microgreens. The farm is visible through a facade-like glass and almost 7 million passengers see it and pass through it every day. Like other forms of underground farms, it uses hydroponic systems in a fully automatic and controlled environment. This vertical farm produces about 30 kg of vegetables daily. The produce is used in the nearby cafes respecting the 'from seed to table' manner (Moon, 2020).

Underground basements of commercial buildings

'Farm.One', which is a vertical farm in Manhattan, grows rare produce for chefs in the middle of the city in a basement under a commercial building. They grow about 500 different micro-greens, herbs and edible flowers and their mission has been to minimise carbon footprint by decreasing shipping. Their rational is that if they can grow rare plants where they are needed and offer them fresh, importing them from far locations would not be necessary. Their business plan is to set up small high-tech farms in different regions so they can serve 'from the farm to the table' (literally to the restaurants in the upper floors). They have even set up a bar that showcases these rare plants and could be visited by guests (Venture City, 2019).

Underground car parks

Since the number of cars in France roads is decreasing and more and more empty underground parking spaces become available, a company started repurposing these abandoned car parks to organic farms. From 2017 to 2020, they sold about 200 Kg of vegetables to Paris grocery stores. One of the reasons they prefer using underground spaces is the thermal stability which protects plants from heat waves and extreme cold throughout the year. The second reason is that parasites and other insects are rarely found in the subsoil (Howes, 2020). La Caverne first started with an underground car park underneath a housing complex with 300 units in Paris. Its inhabitants converted it into a kitchen garden of 3500 cubic meters. They have been able to produce micro-greens, oyster and shiitake mushrooms and different organic vegetables. The fertiliser they use is the compost they make out of coffee ground and their recycled microbrewery dregs (Mancebo, 2017).

Underground spaces far from cities under deserts and mountains

Bunkers under deserts

Many have gone extra miles and set up underground farms in their underground shelters and bunkers far from potential target locations (cities) to ensure in case of any catastrophe such as a severe natural hazard or a nuclear attack, they can still provide their own food underground. The main techniques are the aeroponics and hydroponics using LED technologies for the lighting and considering extra power resources as these bunkers or shelters are often far away from civilisation and run off-grid. Most of the underground food production has been limited to leafy greens when there is light available, but the Vivos Group bunkers actually raise Tilapia too to feed its current residents (Bendix, 2019).

The same company who helped the inhabitants of the building to build La Caverne has made another underground farm this time in 1880s bunker close to the Strasbourg which is called Le Bunker Comestible (Mancebo, 2017). Most underground bunkers are equipped to some form of food production system, particularly those that had prepared for longer stays of 5 years and more. Garrette (2020) brings good examples of different bunkers and their facilities and amenities. Already many companies such as 'Rising S Company' are selling 'grow rooms' and 'greenhouses' with their bunkers listed under their preparedness plans.

Coal mines and tunnels through mountains

UK's coal industry collapsed in 1980s and left many abandoned tunnels and shafts. Advocates believe that subterranean farms can produce 10 times more food and they find using mines and tunnels a very cost-effective way to respond to the food insecurity. Since these tunnels and shafts already exist, they can decrease the costs of vertical farming. The University of Nottingham presented a very interesting concept where plants are suspended and moving in the air and the nutrient and water are pumped and sprayed to their roots. When they get to the top where the harvesting station is situated, they can easily be accessed (Lyod, 2018). The only point to mention is that they will need some kind of force to move the plants and to pump and spray the nutrients. It would be interesting if they consider both electric power and mechanical forces in case of emergencies and integrate with systems to decontaminate the rising groundwater common in many redundant coal mines.

Growing food underground has been opted in some countries including the United Kingdom, the Netherlands, Sweden and South Korea. Doran and Pisa (2019) report a commercial Tunnel Farm which uses a derelict tunnel beneath a mountain in South Korea to produce leafy greens and salads. The tunnel was part of a highway with high risk of collisions which was ultimately closed down. 600 m of the tunnel has converted to a dense stacked-bed vertical farm using pink LED lights and playing Beethoven music to get as much plant as possible. Being inside the ground means keeping a steady temperature and the farm does not need heating which can save a lot of energy.

Discussion and conclusion

Losing calves, wasting 600 millions of dollars of produce including all the citrus, livestock and horticultural crops frozen on trees or in greenhouses after the severe cold wave in Texas (Schattenberg, 2021) should show us that resilience of our agriculture infrastructure is far more important than its intelligence and smartness.

- It is important to remember that due to the aggravation of climate change impacts, the severity and frequency of natural hazards will increase and the chances of getting extreme heat or cold and losing power in the midst of all that will increase too.
- Farmers and food growers should think of mechanical systems as back-up plans to ensure in such cases they do not lose much and still ambient temperature and irrigation can be managed manually if for any reason the automatic systems stop working. They should also keep records of irrigation, pesticide and on non-electricity-dependent devices or mediums like simple old school paper records.
- Ensuring that our food farming produces healthier water and food and less global waste. That means we are closer to achieving these Sustainable Development Goals: Goal 2 (Zero hunger), Goal 6 (Clean Water and Sanitation), Goal 12 (Responsible Consumption and Production), Goal 13 (Climate Action), Goal 14 (Life Below Water) and Goal 15 (Life on Land).
- Urban farms have very important roles to play including '*providing an opportunity to cultivate and produce local produce to those lacking adequate nutrition*' (Siegner et al., 2018, p21, p2).

Now that we are facing serious threats by climate change, mitigating and adapting with climate is our only way forward to prolong the survival of the humankind on the planet.

References

Adewale, C., Reganold, J. P., Higgins, S., Evans, R. D., & Carpenter-Boggs, L. (2019). Agricultural carbon footprint is farm specific: Case study of two organic farms. *Journal of Cleaner Production, 229*, 795–805.

Aerofarms. (2021). *Our indoor vertical farming technology*. Viewed 30 January 2021. Available at https://aerofarms.com/technology/.

Alter, L. (October 11, 2018). *Jargon watch: Vertical gardens vs vertical farms vs living walls vs green Façades*. Treehugger. Sustainability for all. Viewed 18 January 2021. Available at https://www.treehugger.com/jargon-watch-vertical-gardens-vs-vertical-farms-vs-living-walls-vs-green-faaades-4856186.

Andrews, K. (2013). *Pasona urban farm by Kono designs*. Dezeen. Viewed 27 January 2021. Available at https://www.dezeen.com/2013/09/12/pasona-urban-farm-by-kono-designs/.

Ashwanden, C. (2016). *Walipini greenhouses -some DIY tips*. Permaculture Research Institute. Viewed 22 February 2021. Available at https://www.permaculturenews.org/2016/11/28/walipini-greenhouses-diy-tips/.

Avis, R. (2013). *Rob's modified Walipin*. Verge Permaculture. Viewed 22 February 2021. Available at https://vergepermaculture.ca/2013/12/18/robs-modified-walpini/.

Beacham, A. M., Vickers, L. H., & Monaghan, J. M. (2019). Vertical farming: A summary of approaches to growing skywards. *The Journal of Horticultural Science and Biotechnology, 94*(3), 277–283.

Bendix, A. (September 18, 2019). *Survivalists are buying underground doomsday bunkers to prep for the apocalypse. Here's what they look like*. Business Insider. Viewed 16 January 2021. Available at https://www.businessinsider.com/what-makes-a-good-doomsday-shelter-2019-9.

BIGH Farm. (2018). *Ferme Abattoir*. Viewed 30 January 2021. Available at https://bigh.farm/.

Bowery Farming. (2020). *Reimagining farming from the ground up*. Viewed 30 January 2021. Available at https://boweryfarming.com/farms/.

Bozhinova, K. (n.d.). *Gotham Greens Expands to Baltimore*. Foodtank, The think Tank for Food. Viewed at 29 January 2021. Available at https://foodtank.com/news/2018/07/gotham-greens-expands-baltimore-viraj-puri/.

Burnham, S. (2020a). *Using an airport terminal as urban rooftop farm. Reprogramming the city: Adaptive reuse and re-purposing urban objects*. Viewed 26 February 2021. Available at https://reprogrammingthecity.com/using-an-airport-terminal-as-urban-rooftop-farm-jetblues-t5-farm-at-jfk/.

Burnham, S. (2020b). *Growing underground: Urban food in abandoned subway tunnels. Reprogramming the city: Adaptive Reuse and Re-purposing urban objects*. Viewed 26 February 2021. Available at https://reprogrammingthecity.com/growing-underground-urban-food-in-abandoned-subway-tunnels/.

CHCI. (2021). *Canada healthy communities initiative*. Viewed 11 February 2021. Available at https://communityfoundations.ca/initiatives/chci/.

City of Toronto. (2019). *PollinateTO community grants*. Viewed 2 February 2021. Available at https://www.toronto.ca/services-payments/water-environment/environmental-grants-incentives/pollinateto-community-grants/.

CUI. (2019). *The Ontario food terminal: A priceless hub in Ontario's food value chain*. Canadian Urban Institute. Available at https://canurb.org/publications/the-ontario-food-terminal-a-priceless-hub-in-ontarios-food-value-chain/.

Doran, T., & Pisa, K. (December 9, 2019). *This farm is growing food deep beneath South Korean mountains.* CNN. Viewed 21 January 2021. Available at https://www.cnn.com/2019/12/09/asia/south-korea-vertical-farm-intl-c2e/index.html.

Duží, B., Frantál, B., & Rojo, M. S. (2017). The geography of urban agriculture: New trends and challenges. *Moravian Geographical Reports, 25*(3), 130−138.

Eigenbrod, C., & Gruda, N. (2015). Urban vegetable for food security in cities. A Review. *Agronomy for Sustainable Development, 35*(2), 483−498.

EPA. (2021). *Sources of green house gas emissions.* United States, Environmental Protection Agency. Viewed 18 September 2021. Available at https://www.epa.gov/ghgemissions/sources-greenhouse-gas-emissions.

European Confederation of Green Roofs and Walls, (?). *Types of green walls. Green Cities in Europe.* Viewed 18 January 2021. Available at https://efb-greenroof.eu/green-wall-basics/.

Garrette, B. (May 14, 2020). *The bunker builders preparing for doomsday.* BBC Future(Engineering). Viewed 26 February 2020. Available at https://www.bbc.com/future/article/20200513-the-bunker-builders-preparing-for-doomsday.

Goodman, W., & Minner, J. (2019). Will the urban agricultural revolution be vertical and soilless? A case study of controlled environment agriculture in New York city. *Land Use Policy, 83*, 160−173.

Growing Underground. (2015). *Bloomberg article.* Viewed 19 January 2021. Available at http://growing-underground.com/bloomberg-article/.

Hobbs, J. E. (2020). Food supply chains during the COVID-19 pandemic. *Canadian Journal of Agricultural Economics/Revue Canadienne d'agroeconomie, 68*(2), 171−176.

Howes, N. (March 8, 2020). *Organic farms take over abandoned underground parking lots in Paris.* The Weather Network. Viewed 7 April 2021. Available at https://www.theweathernetwork.com/ca/news/article/organic-farms-take-over-abandoned-underground-parking-lots-in-paris-france-cycloponics-mushrooms.

InFarm. (2020). *Celebrating Global G.A.P certification and best farming practices at Infarm.* Viewed February, 2021. Available at https://medium.com/@infarm.com/celebrating-globalg-a-p-certification-and-best-farming-practices-at-infarm-32df222b092b.

Kalantari, F., Mohd Tahir, O., Mahmoudi Lahijani, A., & Kalantari, S. (2017). A review of vertical farming technology: A guide for implementation of building integrated agriculture in cities. In *Advanced Engineering Forum* (Vol. 24, pp. 76−91). Trans Tech Publications Ltd.

Karwacka, M., Ciurzyńska, A., Lenart, A., & Janowicz, M. (2020). Sustainable development in the agri-food sector in terms of the carbon footprint: A review. *Sustainability, 12*(16), 6463.

Lim, T., Baik, Y. K., & Kim, D. D. (2020). Heating performance analysis of an air-to-water heat pump using underground air for greenhouse farming. *Energies, 13*(15), 3863.

Li, X., & Siddique, K. H. (2018). *Future smart food. Rediscovering hidden treasures of neglected and underutilized species for Zero Hunger in Asia, Bangkok.* FAO. Available at http://www.fao.org/home/search/en/?q=Future%20smart%20food.%20Rediscovering%20hidden%20treasures%20of%20neglected%20and%20underutilized%20species%20for%20Zero%20Hunger%20in%20Asia.

Lloyd, M. (December 2, 2018). *Old coal mines can be 'perfect' underground food farms.* BBC News. Viewed 7 April 2021. Available at https://www.bbc.com/news/uk-wales-46221656?intlink_from_url=https://www.bbc.com/news/topics/c52ew8q50z2t/wales-business&link_location=live-reporting-story.

Mancebo, F. (October 28, 2017). *Past and future? Living and growing food underground.* The Nature of Cities. Viewed 26 February 2021. Available at https://www.thenatureofcities.com/2017/10/28/past-future-living-growing-food-underground/.

Moon, K. (July 24, 2020). *Is underground farming the future of food? BBC travel.* Viewed 26 February 2021. Available at http://www.bbc.com/travel/story/20200723-is-underground-farming-the-future-of-food.

Oyuela, A. (2019). *16 Initiatives changing urban agriculture through tech and innovation.* Foodtank, The think Tank for Food. Viewed at 29 January 2021. Available at https://foodtank.com/news/2019/12/16-initiatives-changing-urban-agriculture-through-tech-and-innovation/.

Peters, A. (June 12, 2015). *These vertical farms turn unused city wall space into gardens that grow your lunch.* Fast Company. Viewed 7 April 2021. Available at https://www.fastcompany.com/40503488/this-underground-urban-farm-also-heats-the-building-above-it.

Peters, A. (October 23, 2017). *This underground urban farm also heats the building above it.* Fast Company. Viewed 18 January 2021. Available at https://www.fastcompany.com/40503488/this-underground-urban-farm-also-heats-the-building-above-it.

Pulighe, G., & Lupia, F. (2020). Food first: COVID-19 outbreak and cities lockdown a booster for a wider vision on urban agriculture. *Sustainability, 12*(12), 5012.

Reynolds, L. (November 6, 2019). *Inside one man's quest to grow and forage 100% of his food for an entire year.* Treehugger. Viewed 8 April 2021. Available at https://www.treehugger.com/inside-one-mans-quest-grow-and-forage-his-food-year-4861305.

Rosendo, I. G. (June 15, 2018). *Farming underground in a fight against climate change.* BBC News. Viewed 7 April 2021. Available at https://www.bbc.com/news/business-44398472.

Rotterdam. (2019). *LIFE@Urban roofs.* Viewed 2 February 2021. Available at https://www.rotterdam.nl/english/urban-roofs/.

Sanjuan-Delmás, D., Llorach-Massana, P., Nadal, A., Ercilla-Montserrat, M., Muñoz, P., Montero, J. I., Josa, A., Gabarrell, X., & Rieradevall, J. (2018). Environmental assessment of an integrated rooftop greenhouse for food production in cities. *Journal of Cleaner Production, 177,* 326–337.

Schattenberg, P. (2021). *Agricultural losses from winter storm exceed $600 million.* Texas A&M AgriLife Communications. Available at https://today.tamu.edu/2021/03/02/agricultural-losses-from-winter-storm-exceed-600-million/.

Siegner, A., Sowerwine, J., & Acey, C. (2018). Does urban agriculture improve food security? Examining the nexus of food access and distribution of urban produced foods in the United States: A systematic review. *Sustainability, 10*(9), 2988.

Square Roots, (n.d.). *Next-Gen farm: Tech, farmers, practices, and program.* Viewed 19 January 2021. Available at https://squarerootsgrow.com/categories/next_gen_farm/.

Stimmler-Hall, H. (n.d.). *French startup grows fresh strawberries in recycled shipping containers.* Foodtank, The Think Tank for Food. Viewed 30 January 2021. Available at https://foodtank.com/news/2016/02/french-start-up-grows-fresh-strawberries-in-recycled-shipping-containers/.

Thomas, D. (December 21, 2020). *Supermarkets try to calm food shortage fears amid border chaos.* BBC News Business. Viewed 25 January 2021. Available at https://www.bbc.com/news/business-55393076.

UN. (2018). *68% of the world population projected to live in urban areas by 2050, says UN.* Department of Economic and Social Affairs. Viewed 21 December 2020. Available at https://www.un.org/development/desa/en/news/population/2018-revision-of-world-urbanization-prospects.html.

Varden, (n.d.) *Vertical Garden: Living Wall Systems.* Viewed 18 January 2021. Available at https://www.verticalgardensupply.com/living-walls.

Venture City. (2019). *Why the future of farming is in cities — the big money in vertical farming.* Online video, YouTube. Viewed 19 February 2021. Available at https://youtu.be/LiNI-JUFtsA.

Critical review of smart agri-technology solutions for urban food growing

Negin Minaei[1,2]

[1]*Urban Studies Program, Innis College, University of Toronto, Toronto, ON, Canada;* [2]*CITY Institute, York University, Toronto, ON, Canada*

Abbreviations

AI	Artificial Intelligence
Apps	Applications (mobile and online software)
AR	Augmented Reality
EMSA	European Maritime Safety Agency
EU	European Union
GNSS	Global Navigation Satellite System
GPS	Geographical Positioning System
HD	High Definition
IoT	Internet of Things
LiDAR	Light Detection and Ranging Systems
NASA	National Aeronautics and Space Administration
OXFAM	Oxford Committee for Famine Relief
SDGs	Sustainable Development Goals
UAV	Unmanned Aerial Vehicle
UN	United Nations

Introduction

According to the UN (2020), the second Sustainable Development Goal — Zero Hunger — will not be achieved by 2030, as undernourishment started to increase again from 2015 and by 2030 hunger would surpass 840 million. According to the World Food programme, the COVID-19 pandemic could add 130 million people to the initial 135 million people who suffer from acute hunger by the end of 2020.

Eight mega trends that are the driving force of the need for agri-technology development were identified by Krishnan et al. (2020) and include the following:

(1) Population growth and extra strains on food security and food systems
(2) Rapid urbanisation
(3) Low value addition and low participation in global value chains
(4) Regional trade growth and its competitive advantages
(5) Gender inequity

Digital Agritechnology. https://doi.org/10.1016/B978-0-12-817634-4.00006-9

(6) Lack of nutrition

(7) Climate change consequences and resource scarcity

(8) Growth of agricultural servicification and deindustrialisation threat.

We are witnessing breakthroughs in advanced technologies almost every day. This chapter aims to introduce you some of these smart technologies that have emerged in the food growing systems in cities. Industrial, digital and smart farming technologies for mass food production have been utilised during the past decade and in recent years we observed a variety of smart technologies that aimed to help farming become more sustainable and resilient. These groups of technologies were aimed to be opted by city dwellers to produce food in macro and micro scales. They are often operated digitally and some need WiFi, Bluetooth or the Internet to enable them work using Smart Apps. The IoT-enabled devices and sensors are used to raise yield, monitor weather data and automatically choose the optimum irrigation time, conserve water and control water use and planting times (Snow, 2017). Most of these smart technologies collect data from drones to sensors and apps and the data privacy of farmers stays important.

In this chapter, we review some examples of the smart technologies for urban agriculture and discuss their state of sustainability. Although, technology has progressed incredibly, surprisingly we still lack some crucial and substantial infrastructure that all cities need to have to stay self-sufficient and resilient in case of a catastrophe. Please note that it is not possible to discuss them comprehensively here, but my hope is to direct you to some valuable studies or generate some inspiring ideas or solutions to the existing problems. This chapter aims to avoid very complicated topics such as software development and coding but mentions some studies for a more advanced audience.

COVID-19 and food security issues in cities

COVID-19 pandemic revealed many serious issues in the food production systems in different cities in both advanced and developing countries including the lack of proper food growing and storing infrastructure that can support cities for a considerable amount of time. It was utterly painful to see 500 Ontario dairy farmers were asked to dump 5 million litres of milk per week to keep prices stable and prevent over-supply (BBC, 2020). Zero waste has been the number one priority of Sustainable Food initiative of the Goal 2 (Zero Hunger) from the 17 Sustainable Development Goals of the United Nations. Milk farmers could lower their prices to help the society feel less pressured because of job losses and financial problems due to the COVID-19 pandemic or simply donate the extra milk to the lower income families all across Canada or to countries with higher rates of poverty and lack of food security. Please note that dairy products in Canada are very expensive, for example, 1 L of milk in Canada is sold for 2.48 C$, whilst it's 1.59C$ in the United Kingdom, 1.48C$ in Iran and 1.30C$ in Germany and 1.08C$ in the United States

(Numbeo, 2021). More importantly, farmers not only wasted a high-quality product but also wasted a lot of energy, water and already emitted carbon footprint of producing raw milk without any return. If each litre of raw milk produces a 0.843 Kg of CO_2 (Zhao et al., 2018), 0.46 Kg of N_2O and 26 g of CH_4 (Nonhebel, 2006), then Ontario dairy farmers have emitted 4,215,000 Kg of CO_2, plus 2,300,000 Kg of N_2O and 130,000 kg of methane (CH_4) weekly which shows why despite the general perception of decreasing climate change impacts, the 2020 was the hottest year on records. The financial loss for each 5 million litre is 12,400,000 Canadian dollars and the Government of Canada announced to compensate 11,000 dairy farmers for about 1.75 billion Canadian dollars. Table 9.1 below illustrates carbon footprints and financial consequences of a wrong decision that was made for dairy farmers in Canada in 2020.

The questions are:

- Could we have regional infrastructures that can store raw milk or transform it to dairy products and store it for a longer period of time?
- What kind of technology could be used to ensure cities can store their food in resilient structures and containers for times of distress such as future pandemics or any kind of natural catastrophes or hazards?

With the sanction of Russia, most farmers in the Canada and US are facing a new challenge and that is accessing fertilizers and pesticides. Apparently, with removing Russia from the equation, there will be insufficient amount of crops and oils in addition to higher fees of pesticides and fertilizers which makes farming even more costly. Investing in agricultural research and technology development as well as improving the rural infrastructure by international cooperation illustrates the importance of using smarter ways of producing food for all. Science-based decisions are necessary for a data-driven farming. Since most of the arable lands on the Earth are already farmed, the need to improve productivity of the existing farms is clear.

There is no doubt that Big Data and smart technologies and their applications in the environmental monitoring, data collection and data processing are quite important, but we shall also consider their possible negative impacts on the food production system. Using IoT in agriculture can be challenging due to high uncertainty level of the whole system from weather condition to the natural context on which plants are growing; nevertheless, it is predicted that IoT-based food systems will be the future of agriculture (Verdouw et al., 2019).

Table 9.1 Carbon footprint and financial losses of dumping milk by Ontario dairy in 2020.

	CO_2 (Kg)	CH_4 (g)	N_2O (Kg)	Loss in Canadian dollars
Per litre	0.843	26	0.46	2.48
5,000,000 L	4,215,000	130,000	2,300,000	12,400,000

Source: The author.

Macro scale: smart food growing technologies

Since the audience of this book are mostly familiar with the precision farming, smart farming and digital farming, in this chapter we do not focus in depth on any of them and their definitions. Based on the Bolfe et al. (2020) definition of digital agriculture, most of the examples that I discuss in this chapter are considered digital agriculture technologies. Although, when it comes to AI and AR, the example technologies often belong to the precision agriculture technologies.

Big data, artificial intelligence, augmented reality, smart sensors and citizen science

Artificial Intelligence (AI) has been one of the main smart city technologies. Most industries have found new demand areas to benefit from big data and high-speed computing capabilities. In environmental studies and biology, AI image classifier has been used to identify, classify and assess different images for biodiversity observation (August et al., 2020). According to this research, AI image classifier performs best when a single plant was captured on its wildlife setting. The UK Centre for Ecology and Hydrology has benefited from huge free data sets that were collected by volunteers or from the social media. Dr. Tom August has been one of the researchers who valued the benefits of Citizen Science on studying and monitoring different wildlife from butterflies to plants from images that were sent by volunteers to social media on platforms such as Flickr. Citizen Science has been promoted by the European Commission (2020) and it is defined as a non-professional involvement of volunteers in data collection, data analysis, problem definition, quality assurance and interpretation, in conduct of a scientific research project. It has been particularly very useful for observations of indicator species such as birds and pollinators. Perhaps, Grow Observatory, a University of Dundee's project, has been one of the most successful initiatives that educated thousands of citizens in the United Kingdom and Europe. They taught citizen science participants sustainable practices and environmental monitoring by distributing smart sensors and teaching them planting and the information they need about soil, land and crops (University of Dundee, 2016).

Smart technologies such as apps, mobile Internet and portable sensors have made data collection easy, speedy and even in real-time. With the growth of urban population and food and water demand, the need to design systems and solution to decrease the water consumption is on the rise, particularly because agriculture is responsible for 70% of the world water consumption. It is been predicted that agriculture will face serious water risks and to confront this problem, AI has been identified as a functional helper in precision agriculture and water conservation (Itzhaky, 2021), as it can provide an optimal irrigation mechanism including scheduling and distribution of water according to predicted weather patterns and extracted data from connected sensors and devices. A very good example of using AI in precision agriculture is the development of the affordable smart technology that can help with weed management which decreases costs, environmental pollution, crop damage and pest resistance (Partel et al., 2019). AI helps identifying the weed leaves from the plant leaves and since it's connected to a smart sprayer, the chemicals are sprayed on the target area covered with weeds and not on the plant so less crop contamination occurs. The recent smart technologies that were mentioned in their paper are the 'H-sensors' and 'See and Spray'; both use AI to differentiate weeds from crop plants.

Probably the first time that application of Augmented Reality (AR) was mentioned in agriculture was when Cupial discussed its potential applications in precision agriculture (Huuskonen and Oksanen, 2018). Other examples of recent technologies are a wearable AR technology that can show farmers which parts of their field were treated whilst they are on their tractor or developing a navigation system to enable farming at night times. Researchers suggest AR can help identifying plants, pests and weeds too. The main application of AR is to superimpose information on real-time images of a space. In that sense, some believe AR can simulate and visualise formation, such as growth of a plant and makes managing different farming tasks easier. The educational applications are predictable; for instance, setting up an AR smart garden can help educating students or the public on how plants grow.

Sensors, IoT, irrigation and remote water control

Sensor technology is developing fast. In agriculture industry, the most common sensors are moisture sensors which are planted close to plants in the soil, they track the amount of water and chemicals that goes into the soil and send their data to the clouds to a super computer to analyse and send the results to the irrigation system which generally is a grid of drip tapes (Economist TQ, 2021). Moisture sensors are the most common types of sensors that have been used even by citizen scientists to help collect soil data. They are planted in the soil but can measure more than moisture, often they measure nutrient content and how it changes over time.

IoT stands for Internet of Things, where different devices in a system are all connected to the Internet and can interact with one another in real-time. Some IoT-enabled sensors can actually monitor and measure plant behaviours in addition to the soil and weather data which is a breakthrough in informing incisive irrigation. Other types can measure evapotranspiration which can be enhanced by satellite imagery and weather prediction (Itzhaky, 2021). Perhaps, one of the most important benefits of the IoT technologies is that they can function in full-spectrum and alert farmers as soon as they detect a problem. For example, detecting leaks and malfunctions or drips can be very time-consuming and labour intensive for farmers; some expensive irrigation sensors can find out irregularities and pinpoint the location of the leak and save water and irrigation costs. They are benefiting from access to data points, AI and machine learning.

Tule ET is a hardware device with an evapotranspiration sensor that is installed in a field above the plant canopy and sends the collected data via the Tule's server using a cellular connection. It measures and reports the amount of water in the field, the irrigation application amount, plant stress and recommends the amount of water to apply based on the atmospheric forecast (Tule, 2020). Many similar devices are available in the market. Automated irrigation controls over-watering and under-watering; it prevents crops diseases and wet soil pests and reduces soil erosion. WaterBit is an example of IoT-based solar-powered sensor that is placed under the foliage and measures the moisture. It's data-driven technology which monitors, communicates, automates, analyses and reports the irrigation data to the user. It is easy to use for farmers and enables them to irrigate their farm by using an app (WaterBit, n.d.). Another example is ODO or One Drop One. It has been around since 2015 and is a truly 'smart' and 'sustainable' device which uses solar energy and conserve water considerably. It is a wireless plug and play autonomous multi-valve system. It decreases water consumption using sensors to detect soil moisture, humidity, acidity and ambient light. It can monitor health of plants and can be managed remotely using a mobile app (ODO, 2015).

There are other types of sensors with different applications, like thermal sensors installed in drones for imaging purposes, remote sensing, or motion sensors to track animals at night or wearable sensors. Smart sensors and devices such as wearables are not used by humans only. A Glasgow start-up designed a Smart Collar that tracks cow's health and fertility and alert farmers when they are ready to mate based on monitoring their eating and wandering activities.

Planting and survey technologies: drones and remote sensing

Remote sensing sensors are not new in agri-technology. Perhaps, we can discuss one of the early examples of using GPS sensors in agricultural mobile equipment such as tractors by the John Deere which could benefit farmers and prevent them from making mistakes such as covering the same field twice thus help them save water, fuel and pesticide (Economist QT, 2021).

Paula Marti from the earth observation has been using high-quality satellite imagery (30 by 30 cm), deep learning and algorithms to monitor the forest and vegetation for conservation purposes in different countries. Working with high-quality images means needing super computers which most often are not available, so the common approach that was opted both by the European Maritime Safety Agency and the Environmental Change Institute at the University of Oxford has been to employ the Cloud space and volunteer computers' spaces to run the process. Dr. Frederick Otto reported that they used 30,000 computers all across the globe. They have worked on satellite data to produce reliable models for OXFAM for extreme weather events that could potentially impact food security (Coghlan et al., 2014).

The image data that is collected by drone cameras is also used to recognise healthy vegetation and help farmers to apply the right amount of nitrogen fertilisers

(King, 2017) based on the amount of near-infrared light that is reflected from a patch. This brings us to the next section to discuss drones and their role in the modern agriculture.

Drones definition, technologies and applications

Drone is defined as a remotely controlled unmanned aircraft with its systems including sensors, software, AI and algorithms, motors/actuators, processing unit, wireless networks, memory, energy management and storage and depending on its use, a camera, a holder or other specific parts. It is a type of aircraft that is also called Unmanned Aerial Vehicle which means no human pilot on board is needed. Drones may have embedded computer systems with controllers and chips, high resolutions cameras, GPS technology or radio technology. They can collect variety of data including heat data, images, audio, video, thermal and infrared images and can stream it online at the time (Voss, 2013).

Drones play many roles in the farming process, from spreading seeds in arable lands and helping the large-scale field monitoring to spraying pesticides on target areas with the exact quantity that is needed. Daponte et al. (2019) reviewed drone technologies suitable for precision agriculture to improve productivity explaining more details on the architecture of these types of drones, the way they function and the process of extrapolating information from the images and sensors. I recommend reading their open access paper. Small drones have many limitations including, small payload, low spectral resolution, poor geometric and radiometric performance, low software automation, sensitivity to atmospheric conditions, short flight endurance, possibility of collisions, potential repair and maintenance problems, assistance and funding dependency, safety and security issues, potential social impacts and ethical issues (Paneque-Galvez et al., 2014). Here are two examples: First, Wasp is a bird-size device that has been used by Texas agents to control a situation and check for unseen dangers. Wasp can fly hundreds of feet above the ground and stream videos instantly (Finn, 2011). Second, Parent DJI Phantom 4 that can address most issues of other drones (Incredibles, 2016) such as heavy weight, balance problems, intelligent navigation without dependency on the satellite support. It could improve to have a longer fly up to 5 kilometres with complete control and a live 720p HD view of everything the camera sees. It can visually track any moving object or any selected object and prevent collision by identifying barriers on its way (DJI, 2017).

Puri et al. (2017) summarise the agricultural tasks that drones perform including: (a) agricultural farm analysis by generating 3D maps for soil analysis, (b) careful inspection of every corner of the field which saves time, (c) precision application of water, fertilisers and pesticides in the right area, (d) benefiting from GIS mapping that is integrated in them, (e) monitoring the status of the crops by crop health imaging. In their paper, they review some of the agriculture drones that were in the market by 2017 and describe their applications, forms and specifications in details. Doddamani et al. (2020) count 12 applications for drones in agriculture which complement the former task list including the following: weather monitoring, agri-tourism, estimating yield, optimising seeds, fertiliser and water input and faster reaction to potential threats like weeds, pests and fungi.

The latest application for the drones is to mimic pollinators. Since pollinator numbers are dropping fast, the concern over losing yields has become even more serious. The annual value of wild pollinator for apple, high-bush blueberry, sweet cherry, tart cherry, almond, watermelon, and pumpkin crops has been estimated one and half million dollars in the United States (Bates, 2020). Therefore, many scientists have been looking for alternative solutions. Bubble-blowing drones are the newest devices that were invented to overcome the issue of extinction of pollinators and artificially pollinate fruit trees by blowing pollen-laden bubbles onto blossoms. The experiment showed 95% of the pollinated blossoms on each plant formed fruits (Temming, 2020).

Drone-imaging, GPS and crop mapping software

Since remote sensors can detect physical, chemical and biological elements in far off environments, the appeal for civil applications has increased. Drones are widely used in monitoring environments including woodlands, wetlands, growth and deforestation. They can sense light changes and therefore light pollution in cities. They can identify airborne microorganisms and detect changes in chemical components of atmosphere (Schlag, 2013). Because of drone affordability, flexibility and being equipped with sensors and cameras, they are often used to repeat data collections in local areas to monitor changes, for example, UK Center for Ecology and Hydrology has used Matrice 600, Phantom3 and QuestUAV 300 (August, 2017). Higher resolution data imagery and less expensive drones have been increasingly used in recent years. Agribotix, which is a company specialised in analysing agriculture data, provides software and drones that can map pockets of unhealthy vegetation in relatively huge fields only by using near-infrared images. Using machine learning to train the system recognise weeds, they aimed for an alert system that can inform farmers of growing weeds in their fields. Harper Adam scientists have used surpassed the Agribotix and tested drones that could blast the weeds by laser (King, 2017).

The reason drones are preferred in precision agriculture is that they are better at generating images. The two reasons are the following: First, cloudy weather does not impact their imaging, but satellite-based images are limited to weather conditions. Second, the resolution of images captured by drones is higher than the satellite images; the pixel units are centimetre-level compared to meter-level unit resolution of satellite images. The only downside is that operating drones need some skills and efforts in all stages including pre-flight, flying and image processing (Huuskonen & Oksanen, 2018). They explain comprehensively the ways to use a drone for soil map colour imaging using AR glasses guide, its needed software kit, to develop the kit and automatic segmentation and sampling point selection.

Drones' images combined with ground sensors play a crucial role in precision agriculture. Daponte et al. (2019) explain that semi-autonomous drones are needed in precision agriculture and because these drones fly on a defined flight path with specific aerial routes and flight altitude, they have to be equipped with some form of positioning system or Global Navigation Satellite System such as GPS. Huuskonen and Oksanen (2018) suggest a more accurate positioning system than GNSS and possible methods to set up the user interface. Their choice of drone was DJI Phantom 4Pro which they describe it as the top model amongst small consumer drone models which I mentioned it in the former section. Drones need to have an altimeter as well to enable them fly at constant altitude.

To properly monitor the vegetation in precision agriculture, different types of sensors and cameras are used including multispectral cameras, thermal cameras, RGB camera and Light Detection and Ranging Systems (LiDAR). Some researchers suggest to use different data sets and triangulate them for a clearer vision and more accurate results, but Niemitalo et al. (2021) collected drone aerial image datasets from different sources and had serious challenges to use the data, since the same type of drones were not utilised to capture the images and the cameras, the angles, the logs and the time of shots and therefore the lighting were all different and made working with these data sets very difficult. They have a good list of recommendations on what to avoid to ensure your configuration and data analysis is as accurate as possible.

Nature conservation, environmental reporting and monitoring

Monitoring the application and implementation of environmental-related policies and regulations is as important as monitoring nature and environment for probable changes. Most of the data is initially collected by locals and are reported by the public. EU legislation requires its members to report particular data and after ensuring that the collected data by the public is reliable and can help with reporting and providing evidence, they encourage citizen science initiatives (European Commission, 2020). They published a full report on the best practices of environmental monitoring by citizen science in 2020. Application of drones in conservation generally fits in two categories of research and direct conservation. According to Paneque-Galvez et al. (2014), small drones have been used in many environmental monitoring researches including biodiversity, habitat monitoring, soil properties, mapping and monitoring fires, poaching and in agriculture. Few academic studies found drones being used in forestry; Paneque-Galvez et al. state that drones have been used for Community-Based Forest Monitoring Programs in Tropical Forests for the first time and suggest that application can reduce tropical deforestation and help the climate change mitigation. Krupnick and Sutherland suggest using drones for forest restoration, to employ drones to deliver seeds; but it is mainly used by law enforcement to monitor illegal activities such as hunting wildlife, deforestation and locating perpetrators (Sandbrook, 2015). However, villagers in Myanmar have started planting mangrove trees with the help of Biocarbon Engineering to restore the local forests to plant 100,000 trees in a single day (Lofgren, 2017). Non-academic literature, however, illustrates that timber companies and government forestry agencies use drones to document tree grown/gap maps, to estimate the volume, to assess the wind blow, monitor pests and plan for harvest. Monitoring biological features such as woodlands and observing, counting and protecting wildlife can provide data for measuring forests biodiversity and for conservation (He, 2015; Sandbrook, 2015) which all are important tasks done by drones. For example, in Namibia as part of the wildlife protection, MicroMappers were used to capture aerial images of semi-arid savanna. The team used the crowd-sourcing technique to identify wildlife in images by sending the images to a remote team of volunteers who would click on the damaged locations after analysing those drone images. Larger number of clicks indicated a drone investigation was needed. Their results showed 87% accuracy (iRevolutions, 2014).

Environmental impacts

Drones are generally considered environmentally friendly as they do not pollute the air and only consume electricity (He, 2015). Noise emission and local ecologic impacts of drones particularly on birds have been mentioned as key problems of drones (Kunze, 2016). Surely, this could vary depending on the types of drone (energy use, environmental footprints, environmental burden), their applications (postal delivery, shopping, medication delivery …) and the distance or area they cover during each flight. The important point to think about is all these devices use energy, they have batteries, they are made of materials and they use resources, which means ultimately, they add more to the non-recyclable waste and that is not sustainable.

Santangeli et al. (2020) wants to turn this around and use drones to benefit the environment and particularly to protect biodiversity amongst birds by adding thermal sensors to drone to capture thermal images. Their study showed that with a semi-automated system that uses both thermal images captured by drones and AI and 'deep learning', it is in fact possible to identify bird nests' locations on the ground before and after a land is gone through farming processes which is a good help to farmers. It can be on an arable land but the type of field, weather and the height of the drone can impact the investigation to some degree.

Micro scale: food growing technologies for homes

Since the concept of sustainable food growing has found favour amongst people who like to help the environment and grow food in cities, a series of similar growing systems have been introduced to the market as self-sustaining low maintenance devices that claim to produce herbs year round at home. Almost all of them are unsustainable products which not only consume electricity to work for their lighting system or pumps but also they eventually end up as non-recyclable items to the solid waste bins and landfills. Here are some examples from this group of products that have been in the market since 2016:

Growing devices

- Click and Grow (2020) is a Smart herb garden to grow three plants in nano-material pots that have a good access to oxygen and nitrogen and can grow plants without watering and fertilizing them. This technology was inspired by a NASA technology.
- Lilo is a self-sustaining indoor garden which grows fresh herbs and mini-vegetables on kitchen counters all year round. It has three biodegradable pods with all the necessary nutrients and the seeds in them and is equipped with the LED light above them that simulates sunlight. After 30 days, the harvest is ready for 3 months but later they would need to discard this pod and buy a new pod to grow another plant (Lilo, 2021) which is not sustainable.
- Tableau, the automatic plant watering tray, is another device which irrigates plants automatically inspired by the wet-dry cycles of nature and is listed under smart growing devices (Uncrate, 2021).
- Ogarden Smart (2021) which is advertised as a Smart Indoor Garden allows you to grow 90 herbs or vegetables at the same time. This is the latest product. It talks about saving money on fresh produce and having vegetables all year round. It is on the market, but like former ones, some seed pots are provided to be put in the germination area; no soil is provided and people will need to buy the pots for each round of planting. Plants will be rotating gently to provide air movement. It has artificial light, and moisture sensors to ensure plants are watered when they need it. The device will inform you when it's time to fill the water tank. It does not explain when they grow more than the diameter of the space and how should they be treated. The device is sold for about $1400 Canadian dollars and 90 indi-vidually wrapped cups of soil for $80. Please note a bag of 15 kg high-quality gardening soil is sold for $15. So it consumes electricity and is semi-automatic.
- There are many examples of micro aquaponic food growing systems which have an aquarium and a vegetable growing section above it topped up by a grow light.

Composting devices

From one hand, global waste is expected to increase to 3.40 billion tonnes by 2050 and whilst 44% of the produced waste is food and green, only 5.5% of it is com-posted; the rest is often dumped openly on the landfills (The World Bank, 2021). From the other hand, healthy topsoil that is a fundamental component to our food system is rapidly degrading and that causes serious problems to grow enough nutri-tious food for the fast growing population (Cosier, 2019). In order to solve the two problems of accumulating waste and losing healthy soils, most smart or sustainable cities such as Toronto started collecting organic waste and producing compost for almost a decade, whilst some unsustainable cities such as Windsor (ON) still does not provide any organic waste collections, only solid waste and recycling waste.

Industry has identified the problem and smart bins are hugely advertised in Smart Waste Management systems for the Smart City concept, as they are equipped with sensors and are IoT-enabled. They can compact waste to allow a bin accommodate a bigger volume of garbage upon being fully filled when they send an alert to the control panel for a pick-up which eventually saves fuels and decreases unnecessary trips and air pollution. The Ecube Labs designed a wireless ultrasonic fill-level sensor that can monitor the trash level and sends real-time data to the network (Ecube Labs, 2021). Whilst smart bins are trying to incorporate solar panels to save energy, the smart organic recycling bins often use more electricity to compost the organic waste specifically from the kitchen waste. One of the most advertised products to environmentally conscious people is composting devices to turn the kitchen food waste into soil. They have come with variety of names from 'Smart compost' to 'intelligent bin', 'automatic garbage recycler', to 'self-sealing and self-changing trash can' and 'sensored trash can'. Nearly all of these smart devices work with electricity and consume power for hours. The earliest or perhaps the first product in this category that I found in 2017 was Zera, an award-winning technology that could compost the kitchen food scraps into fertilisers in 24 hours (WLabs, 2021). It has won numerous awards for innovation and excellence. Since then, similar technologies have been advertised in the market. The main selling points for this product and similar ones are their user-friendly applications, they reduce food waste volume by over two-thirds its original volume (Zera claims by 95%), they are odourless, pestless and are fully automated which makes customers more willing to opt for them rather than mechanically producing compost in their backyard or balcony that takes a long time and a lot of man power and possibly bad smell and insects. With labels such as Zero-Waste, more people are encouraged to do their share to protect the planet but only those who pay attention to the details see that some products work 24/7 consuming electricity which clearly shows they are unsustainable solutions. For example, Earth Matters (2021) introduces two types of products for both mesophilic and thermophilic composting which most of them work mechanically with a handle and does not need electricity. Composting aims to decrease the amount of organic waste and transform them into some natural fertiliser that can benefit plants not to produce devices that are not recyclable and add more solid wastes to the landfill. As most of the designed composting machines are made of plastic and in case of being smart, they are equipped with sensors and chips and wires.

My MEng students were able to design and prototype real sustainable composters in the Sustainable Smart Cities course. The aim was a type of composter that did not have any negative environmental impacts. Some companies advertise more sustainable options. There are good technological solutions that do not use any electricity at all but simultaneously produce energy. One good example is the Home Biogas (2020) which allows families to compost their food wastes and produce their own cooking gas and fertilisers from it.

Smart food storage devices

As explained in the first part of this chapter, lack of resilient storing infrastructure is still the biggest problem for the times of crisis which causes considerable amount of food waste, emissions and costs for both farmers and governments. Waldman et al. (2020) have studied resilient food storage and propose by collecting weekly food storage data from households and their purchases; small farm holders are able to find the relation between grains and food, patterns of demand and storage decline, food security and food insecurity and therefore invest on a proper food storage system. A review study by Terrascope, Mission 2014 (n.d.) at the MIT university recognises the two most efficient and economical ways of storing food: metal silos to store grains and solar dryers to store meat, vegetables and fruits. It is important because not only it adds to the food safety just right after the harvest, it also prevents food waste which apparently has a significant share after harvest. These are the best solutions that farmers could use, but what about domestic use? How can we safely store our food for a longer time to prevent food waste? You guessed it right, new patented inventions using smart technologies, connected devices and apps have come to our kitchens to solve this problem. But are they sustainable solutions or merely trendy gadgets on kitchen counters? Do they really solve an unsolvable problem without causing environmental burdens and consuming energy?

Silo is a countertop deep vacuum system that with a press of a button can create an airtight storage space for your vegetables and fruits and inform you when your food is gone bad. It is equipped with freshness sensors and of course with Wifi and a voice control system. You can ask Alexa whether your food is still fresh and edible. As you can imagine like any other IoT-enabled device, it is connected to the Clouds and is run by an app; it can track your consumption and inventory. This device was designed and patented by Silo which is a Tel Aviv-based start-up. The container is BPA-free. According to the producer, food can last five times longer than normal in this device (The Kitchen Guy, 2018). It has limited capacity so if you need more you would need to buy more countertop devices. The question that crosses my mind is which option lasts longer, storing fresh vegetables in air-tight containers and leaving them in the fridge or leaving them in this device on the countertop without any cooling mechanisms.

Another recent 'Smart' gadget which is not really smart is called Ovie Smarter-Ware (2019). It helps you to keep track record of your food, particularly when you stored them by adding a smart LED tag and a smart lid to the container. One can simply label the food with stickers instead of buying the whole set of smart tags, a Hub, a smart phone, an app and a smart speaker such as Alexa to record and send all the information on the Clouds and inform when the food is not safe to eat! Another product to add non-recyclable waste to the solid waste after its lifecycle is over. If we are wondering how humans could produce so much information compared to former decades, here is an example of the type of information generated and stored in the Clouds and the Internet on a daily basis.

PantryChic adds more benefits to the Silo style storage food. It is the combination of an electric kitchen scale with different airtight containers. It can measure the ingredients you store in a container and put it on top of the base and allow you to dispense the exact amount you need (Lamb, 2018). By using an app it informs you when you are running low of an ingredient or helps you to find recipes using those ingredients. As always, electricity and connectivity are needed to enable the device to work.

Stasher produces reusable silicone bags that can store both food or used in sous vide cooking technique. Since this technique of cooking meat is gaining popularity and at the same time people do not want to use single-use plastic (Lamb, 2018), the marketing idea seems to have attracted enough attention. They all claim that they are ending the single use plastic consumption, whilst no one talks about the cross-contamination of using reusable bags. A study by Barbosa et al. (2019) showed that the risk to health due to cross-contamination is real, as they found high number of multi-resistant Enterobacteriaceae and Staphylococci in addition to *Listeria Monocytogenes* in the bags. Also, I wonder if cooking meat or any kind of food with sous vide method in plastic bags can cause risks to peoples' health? I can imagine any kind of fish or meat already contain micro-plastics, what if putting them a plastic bag in hot water releases more micro-plastics migrating to the meat inside of the bag? I could not find a peer-reviewed published paper or a direct answer but could find a short study which was after the same question (PPRC, n.d.), and due to lack of data on that, that study could not definitively answer as well.

I have recently come across solutions that most preppers are familiar with including freeze drying, canning, and using vaccumed Mylar bags (air-free) to store food for a long time. In terms of freeze drying the claim is that the food could be stored safely up to 25 years. The other interesting storage technique could be learned from the so-called MRE (Meals Ready to Eat) foods which have a shelf-life of about 5 years and they are more known to military and preppers. Radio-wave Dryer is another option to help store food for a longer period of time but can it help to store fresh produce in large scale in emergency situations like pandemics and hazards? What kind of structures can store milk in case of another power shut down or another pandemic?

Discussion and conclusion

Not all farmers are tech savvy and it is important to make research, data and technology accessible to them by raising awareness and providing educational and consultation services as this is the one important step to help them optimise their food production, save water and energy and decrease their wastes and costs. Whilst all these advanced technologies can benefit food production processes, increase the volume, decrease environmental footprints and prevent water waste, there are some facts that we should have in mind:

- *'All data technologies including computers, mobile phones, Big Data, Artificial Intelligence, IoT and Smart City technologies work with and depend on electricity'* (Minaei, 2021). Power generation can be challenging at times. So relying merely on electricity may not be a reasonable option. The recent experience of farmers in Texas after the snow storm with its devastation affects their operation such as daily dumping of $8 million litres of milk because of lack of power in their plants (Wallis, 2021) should illustrate solidly that we need to think to mechanical backup systems.

- Farmers should not rely merely on Machine Learning, Clouds and advanced technologies. They shall keep in mind that all these systems work whilst the Internet is working and literally whilst the power is on. Simple glitches in a power grid, cyber attacks, EMP (Electromagnetic Pulses) or natural hazards can cut the power and all Agri-Technologies that depend on electricity can shut down at once. Thinking of backup systems is essential particularly for bigger scale farms and food production industries.
- Farmers should know that all Internet and Cloud systems can be prone to hacks, and therefore, thinking of data security mechanisms is essential. It is always a good idea to ensure they can overrule the system in cases of hacks and breaches and I am not even mentioning virus attacks and AI.
- Whilst macro-enterprises can advance their food production, micro-enterprises and small farm farmers are less likely to benefit from the Smart Agri-Technologies due to the three difficulties of cost, complexity and capabilities (Krishnan et al., 2020).

When it comes to technology, we should ensure our investments, innovations and designs are all in compliance with the resilience and sustainability goals for the planet. It is wise to plan for the worst case scenario, so in case of better scenarios, we are still ready to face the challenges that are coming at us. At the end of my former chapter titled 'Critical Review of Urban Energy Solutions and Practices' for a book titled '*Sustainable Engineering for Life Tomorrow*' Minaei (2021), after I carefully looked at some food growing and food recycling technologies, I proposed a list of questions that I thought if inventors asked themselves in different phases of design — from idea to product — it could help them see whether their design was truly sustainable or not. Eventually, we only have one available planet to live on and must conserve and protect it.

References

August, T. A. (February 28, 2017). *Science takes to the skies with 'Hollywood-style' remote sensing drone technology*. UK Center for Ecology and Hydrology. Viewed 4th January 2021. Available at https://www.ceh.ac.uk/news-and-media/blogs/science-takes-skies-hollywood-style-remote-sensing-drone-technology.

August, T. A., Pescott, O. L., Joly, A., & Bonnet, P. (2020). AI naturalists might hold the key to unlocking biodiversity data in social media imagery. *Patterns, 1*(7), 100116. https://doi.org/10.1016/j.patter.2020.100116

Barbosa, J., Albano, H., Silva, C. P., & Teixeira, P. (2019). Microbiological contamination of reusable plastic bags for food transportation. *Food Control, 99*, 158—163.

Bates, T. (August 11, 2020). *Bee population numbers are dropping—and US crop yields could plummet as a result*. World Economic Forum. Viewed 18th Jan 2021. Available at https://www.weforum.org/agenda/2020/08/loss-of-bees-threatens-us-crop-yields/?fbclid=IwAR2G7Dt7MMEw2FRG1EZ_7WEjrbbubhke9JY0wifQ3TRUwO3C9lQL6QFWdSI.

BBC. (April 6, 2020). *Coronavirus: Why Canada dairy farmers are dumping milk*. Viewed 25th Jan 2021, Available at https://www.bbc.com/news/world-us-canada-52192190.

Bolfe, É. L., Jorge, L. A. D. C., Sanches, I. D. A., Luchiari Júnior, A., da Costa, C. C., Victoria, D. D. C., Inamasu, R. Y., Grego, C. R., Ferreira, V. R., & Ramirez, A. R. (2020). Precision and digital agriculture: Adoption of technologies and perception of Brazilian farmers. *Agriculture, 10*(12), 653. https://doi.org/10.3390/agriculture10120653

Click and Grow. (2020). *Meet the smart garden - the gift that keeps on giving*. Viewed 21st December 2020. Available at https://www.clickandgrow.com/.

Coghlan, C., Muzammil, M., Ingram, J., Vervoort, J., Otto, F., & James, R. (2014). *A sign of things to come? Examining four major climate-related disasters, 2010–2013, and their impacts on food security.*

Cosier, S. (May 30, 2019). *The world needs topsoil to grow 95% of its food — but it's rapidly disappearing*. The Gaurdian. Viewed 27th Jan 2021. Available at https://www.theguardian.com/us-news/2019/may/30/topsoil-farming-agriculture-food-toxic-america.

Daponte, P., De Vito, L., Glielmo, L., Iannelli, L., Liuzza, D., Picariello, F., & Silano, G. (2019). May. A review on the use of drones for precision agriculture. In *IOP conference series: Earth and environmental science* (Vol. 275, No. 1, p. 012022). IOP Publishing.

DJI. (2017). *Phantom 4*. Vised 15 March 2017. Available at https://www.dji.com/phantom-4.

Doddamani, A., Kouser, S., & Ramya, V. (2020). Role of drones in modern agricultural applications. *Current Journal of Applied Science and Technology*, 216–224.

Earth Matters. (2021). *Methods and compost devices*. Viewed 21 Jan 2021. Available at https://earthmatter.org/compost-learning-center/methods-and-compost-devices/.

Economist TQ. (2021). *Technology quarterly; the future of agriculture?*. Viewed 15 Jan 2021. Available at https://www.economist.com/technology-quarterly/2016-06-09/factory-fresh.

ECube Labs. (2021). *CleanCUBE, the solar-powered trash compactor*. Viewed 26 Jan 2021. Available at https://www.ecubelabs.com/.

European Commission. (February 27, 2020). *Best practices in citizen science for environmental monitoring (SWD), Brussels*. Viewed 15 Jan 2021. Available at https://ec.europa.eu/jrc/communities/en/community/examining-use-and-practices-citizen-science-eu-policies/page/best-practices-citizen.

Finn, P. (January 23, 2011). *Domestic use of aerial drones by law enforcement likely to prompt privacy debate*. The Washington Post. Viewed 8th February 2017. Available at http://www.washingtonpost.com/wp-dyn/content/article/2011/01/22/AR2011012204111_pf.html.

He, Z. (2015). August. External environment analysis of commercial-use drones. In *2015-1st International symposium on social science*. Atlantis Press.

Home Biogas. (2020). *The machine that converts your waste into clean energy*. Viewed 27 March 2020. Available at: https://www.homebiogas.com/.

Huuskonen, J., & Oksanen, T. (2018). Soil sampling with drones and augmented reality in precision agriculture. *Computers and Electronics in Agriculture, 154*, 25–35.

Incredibles. (March 2, 2016). *Top best drones available*. Online video. Viewed March 2017. Available at https://www.youtube.com/watch?v=4tm12YI6FQ8.

iRevolutions. (September 9, 2014). *Piloting microMappers: Crowdsourcing the analysis of UAV imagery for disaster response*. iRevolutions. Viewed 28 March 2017. Available at https://irevolutions.org/2014/09/09/piloting-micromappers-in-namibia/.

Itzhaky, R. (January 5, 2021). *How AI will solve agriculture's water efficiency problems*. World Economic Forum. Viewed 10 Feb 2021. Available at: https://www.weforum.org/agenda/2021/01/ai-agriculture-water-irrigation-farming/.

King, A. (2017). Technology: The future of agriculture. *Nature, 544*, S21–S23. https://doi.org/10.1038/544S21a

Krishnan, A., Banga, K., & Mendez-Parra, M. (2020). *Disruptive technologies in agricultural value chains. Insights from East Africa*. Working paper 576.

Kunze, O. (2016). Replicators, ground drones and crowd logistics a vision of urban logistics in the year 2030. *Transportation Research Procedia, 19*, 286–299.

Lamb, C. (October 2018). *Tupperware is fine and all, but A new wave of Smart food storage is here*. The Spoon. Viewed 9th April 2021. Available at https://thespoon.tech/tupperware-is-fine-and-all-but-a-new-wave-of-Smart-food-storage-is-here/.

Lilo. (2021). *Lilo, A garden right at your fingertips*. Viewed Jan 2021. Available at https://pretapousser.co.uk/indoor-gardens/lilo-indoor-herb-garden-18821.html.

Lofgren, K. (2017). Drones are planting an entire forest from the sky. In *Inhabitat - green design, innovation, architecture, green building*. Viewed 14 August 2017. Available at http://inhabitat.com/drones-are-planting-an-entire-forest-from-the-sky/.

Minaei, N. (2021). Chapter 3 titled: "A critical review of urban energy solutions and practices". In Stagner, & Ting (Eds.), *Sustainable engineering for life tomorrow*. Lexington Books/Series: Environment and Society.

Niemitalo, O., Koskinen, E., Hyväluoma, J., Tahvonen, O., Lientola, E., Lindberg, H., Koskela, O., & Kunttu, I. (2021). A year acquiring and publishing drone aerial images in research on agriculture, forestry, and private urban gardens. *Technology Innovation Management Review, 11*(2), 5–16.

Nonhebel, S. (2006). Options and trade-offs: Reducing greenhouse gas emissions from food production systems. In *Agriculture and climate beyond 2015* (pp. 211–230). Dordrecht: Springer.

Numbeo. (2021). *Price rankings by country of milk (regular), (1 liter) (markets)*. Viewed at 25th January 2021. Available at https://www.numbeo.com/cost-of-living/country_price_rankings?itemId=8&displayCurrency=CAD.

ODO. (2015). *One Drop One solutions*. Viewed at 27th January 2021. Available at https://www.onedropone.net/.

OGarden Smart. (2021). *What is an OGarden Smart?* Viewed at 15 April 2021. Available at https://ogardensmart.com/.

Ovie. (2019). *Keep tabs on your food like never before*. Viewed 9 April 2021. Available at https://ovie.life/.

Paneque-Gálvez, J., McCall, M. K., Napoletano, B. M., Wich, S. A., & Koh, L. P. (2014). Small drones for community-based forest monitoring: An assessment of their feasibility and potential in tropical areas. *Forests, N, 5*, 1481–1507. https://doi.org/10.3390/f5061481

Partel, V., Kakarla, S. C., & Ampatzidis, Y. (2019). Development and evaluation of a low-cost and Smart technology for precision weed management utilizing artificial intelligence. *Computers and Electronics in Agriculture, 157*, 339–350.

PPRC, (n.d.). Do plastic chemicals leach into food from Sous Vide (SV) Cooking? Viewed 9th April 2021. Available at https://pprc.org/2013/p2-rapid/do-plastic-chemicals-leach-into-food-from-sous-vide-sv-cooking/.

Puri, V., Nayyar, A., & Raja, L. (2017). Agriculture drones: A modern breakthrough in precision agriculture. *Journal of Statistics and Management Systems, 20*(4), 507−518.

Sandbrook, C. (2015). The social implications of using drones for biodiversity conservation. *Ambio, 44*(Suppl. 4), S636−S647.

Santangeli, A., Chen, Y., Kluen, E., Chirumamilla, R., Tiainen, J., & Loehr, J. (2020). Integrating drone-borne thermal imaging with artificial intelligence to locate bird nests on agricultural land. *Scientific Reports, 10*(1), 1−8.

Schlag, C. (2013). The new privacy battle: How the expanding use of drones continues to erode our concept of privacy and privacy rights. *Journal of Technology Law & Policy, XIII.* https://doi.org/10.5195/tlp.2013.123

Snow, J. (April 13, 2017). *From drone surveys to Smart tractors, agriculture goes high-tech.* Smart Cities Dive. Viewed 21st December 2020. Available at https://www.smartcitiesdive.com/news/from-drone-surveys-to-Smart-tractors-agriculture-goes-high-tech/440389/.

Temming, M. (June 22, 2020). *Bubble-blowing drones may one day aid artificial pollination.* ScienceNews. Viewed 18th January 2021. Available at https://www.sciencenews.org/article/bubble-blowing-drones-may-one-day-aid-artificial-pollination.

Terrascope, Mission 2014 (n.d.). Feeding the world: Food storage system: Solar dryers & metal silos. Scripts. MIT.edu. Viewed 9 April 2021. Available at http://12.000.scripts.mit.edu/mission2014/solutions/food-storage-system-solar-dryers-metal-silos.

The Kitchen Guy. (October 16, 2018). *Silo: The food storage solution.* Medium. Viewed 7 April 2021. Available at https://medium.com/@thekitchenguy_/silo-the-food-storage-solution-3e3f63719c30.

The World Bank. (2021). *What a waste 2.0, A global snapshor of solid waste management to 2050. The World Bank website, trends in solid waste management.* Viewed 27th Jan 2021. Available at: https://datatopics.worldbank.org/what-a-waste/trends_in_solid_waste_management.html.

Tule. (2020). *Tule vision.* Viewed 21st December 2020. Available at https://www.tuletechnologies.com/.

UN. (2020). *Goal 2: Zero hunger. Sustainable development goals.* Viewed 21st December 2020. Available at https://www.un.org/sustainabledevelopment/hunger/.

Uncrate. (2021). *Tableau automatic plant watering tray.* Viewed 27th Jan 2021. Available at https://uncrate.com/tableau-automatic-plant-watering-tray/.

University of Dundee. (2016). *Public engagement project: Grow observatory.* Viewed 15 Jan 2021. Available at https://www.dundee.ac.uk/projects/grow-observatory-0.

Verdouw, C., Sundmaeker, H., Tekinerdogan, B., Conzon, D., & Montanaro, T. (2019). Architecture framework of IoT-based food and farm systems: A multiple case study. *Computers and Electronics in Agriculture, 165*, 104939.

Voss, W. G. (2013). Privacy law implications of the use of drones for security and justice purposes. *International Journal of Liability and Scientific Enquiry, 6*(4), 171−192.

Waldman, K. B., Giroux, S., Blekking, J. P., Baylis, K., & Evans, T. P. (2020). Smallholder food storage dynamics and resilience. *Food Security, 12*(1), 7−20.

Wallis, J. (February 16, 2021). *Texas agriculture commissioner issues 'red alert' for food supply chain.* KHOU11. Viewed 19 February 2021. Available at https://www.khou.com/

article/weather/texas-agriculture-commissioner-issues-red-alert-asking-gov-abbott-for-help/287-7980b04b-3b63-4cc5-a08e-8a952501bfc5.

Water Bit, (n.d.). Achieve maximum quality and target yield while optimizing labor and water use. Viewed 21st December 2020. Available at https://www.waterbit.app/.

WLabs. (2021). *Food scraps today, fertilizers tomorrow.* Viewed at 27th Jan 2021. Available at https://wlabsinnovations.com/pages/zera.

Zhao, R., Xu, Y., Wen, X., Zhang, N., & Cai, J. (2018). Carbon footprint assessment for a local branded pure milk product: A lifecycle based approach. *Food Science and Technology, 38*(1), 98−105.

Agriculture 4.0: data platforms in food supply

10

Sinead Quealy[1], Patrick Joseph Lynch[2], Narjis Hasan[1]

[1]*VirtualVet, Kilmacthomas, United Kingdom;* [2]*RIKON Research Centre Waterford Institute of Technology, Waterford, Ireland*

Introduction

Approximately 12,000 years ago, agriculture triggered the "Neolithic Revolution" or the "First Agricultural Revolution" when the traditional hunter-gatherer lifestyle was replaced with permanent settlements and animal domestication, and we assume a reliable food supply chain visible to the local consumers. This innovation in agriculture spurred urban development, the growth of civilization, and the rapid expansion of the global population from some five million people 10,000 years ago, to around two billion at the start of the industrial revolution. Subsequent agricultural revolutions can be characterized as the application of basic science to farm management Agriculture 2.0. The application of synthetic chemicals accelerated plant breeding and power machinery as Agriculture 3.0 in the twentieth century and now the digital revolution, which has been labeled as Agriculture 4.0 (Fig. 10).

The evolution of agriculture 4.0

In the last 10,000 years, agriculture has evolved through several transformations from Agriculture 1.0—4.0, as shown in Fig. 10.1. Agriculture 1.0, also called the First Agricultural Revolution, refers to the traditional agricultural era. It was characterized by stationary farming and relied on manual labor, horse power, and simple tools, which meant that productivity remained relatively low. Starting in the 17th century and ending in the 19th, the second revolution or Agriculture 2.0 took place when European agriculture and its diaspora in the Americas, Australasia, and parts of Africa shifted from the techniques of the past. Farming changed dramatically in this period with the introduction of crop rotations and drainage, which greatly increased crop and livestock yields by improving soil fertility and reducing fallow, the mechanization of activities with the invention of new horse drawn tools, such as the plow, thrashing machine, and seed drill, which greatly improved the efficiency of operations. Innovations such as railroads and intercontinental steam ships. From the early 1900s, the use of synthetic fertilizer, feed concentrate, and machinery resulted in not only more food being produced but that food being transported more easily. Increased production from decreased labor demands meant migration to cities and urban expansion.

Digital Agritechnology. https://doi.org/10.1016/B978-0-12-817634-4.00011-2

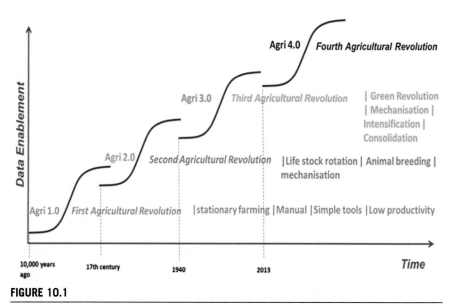

FIGURE 10.1

The history of agriculture can be characterized as a series of waves of innovation creating revolutions in productivity. In Agriculture 4.0, the importance of data aggregation and analysis from multiple sources has become a vital component.

Between 1916 and the late 1960s, Agriculture 3.0 or the Third Agricultural Revolution emerged by the application of the internal combustion engine and rural electrification, better plant and animal breeding, and chemical fertilizer and pesticides. Such was the increased food production around the world at this time; this revolution is also called the Green Revolution powered by the introduction of biotechnology, genetic engineering and advancements in technologies, alongside computerized programs robotics, and heavy agricultural machineries all, more efficient operations. As is now accepted, this phase is identified with an over-reliance on chemicals along with mechanical degradation of the soil.

Agriculture 3.0 helped to establish intensive agricultural production methods globally, and its legacy of intensification, consolidation, and specialization of agricultural practices during this era can still be seen today, whereby a small number of actors control or influence most of the agriculture value chains from production to processing to retail. With government backing, this period is predominantly characterized by promoting technology as a means of sustainably intensifying agriculture to increase productivity and yields (Dicks et al., 2019). Notwithstanding the obvious negative impacts of this productivist lens on the natural environment and the sustainability of the rural economy, a recognition emerged that while big machines and chemicals worked for a long time, they cannot be used sustainably in the same way in the future, and the innovation cycle became limited. As with each revolution, the consumer became more and more separated from where food was produced and processed, and methods of guaranteeing and reporting provenance became a requirement to prevent fraud and reassure customers.

The main focus of Industry 4.0 is to transition industrial production toward a digital model in order to increase efficiency and productivity by mainly focusing on optimized production processes, reduced development time and cost, and enhanced customer value. Much like the Industry 4.0 focus, the fourth agricultural revolution refers to the employment of technologies such as the Internet of Things, Big Data, artificial intelligence, cloud computing, remote sensing, etc. for the generation and processing of big data as a foundation for decision-making. Because it is at an early stage of development, there are a lot of different terms used by policy makers, academics, and practitioners that refer to the fourth agricultural revolution including Agriculture 4.0 or Agri 4.0. Agriculture 4.0 involves the application of these technologies for advanced analytics of large amounts of data to improve the efficiency of agricultural activities. Agriculture 4.0 is distinct from its forerunners and may have the potential to transform agriculture and overcome the challenges facing the industry. It promises to do so not just by increasing the global food production through improved use of data but also by enabling adaptions to climate change and reduce the food loss and waste in the supply chain system. Better control should enable agriculture to use natural resources in a sustainable way and also provide ecosystem services, biodiversity, clean water, and air.

The twenty-first-century global agriculture value chain is being challenged as never before, and it now requires a new revolutionary shift triggering a change in society and the way in which people live. Over the next 30 years, the global population is expected to increase by 2 billion persons from today's levels, reaching 9.7 billion by 2050, and is expected to reach its peak around the end of the current century at a level of nearly 11 billion (2019 Revision of World Population Prospects). With the increasing population in the world, the expected demand for agricultural products will become higher than ever before adding pressure on the agriculture supply chain. Indeed, according to the Food and Agriculture Organisation of the United Nations (FAO), the global level of agricultural production and consumption in 2050 is projected to be 60% higher than today. Long distance food supply has been a feature of all major civilizations and has increased dramatically in the industrial age only a few impoverished countries practice autarchy in the modern age.

However, reaching this level of production to feed a burgeoning global population has a number of challenges. First, the demand for agricultural products has been historically achieved by increasing the amount of cultivated land use on the planet, which currently accounts for approx. 38% (Rizvi et al., 2018). It is probably safe to say that this is not environmentally or economically sustainable, especially given that to meet the growing global food demand would require approximately 320–850 million hectares of agricultural land additionally by 2050ii. It is interesting to note that of the remaining global land surface approximately 30% is unsuitable for cultivation on account of soil, climate, and urban development. Moreover, the ecological and social trade-offs of clearing more of the remaining 32% of the global land surface for agriculture are often high particularly in the tropics, where the land is covered in natural land states such as forests. So, very little land is available for agriculture expansion. Second, changing consumer dietary habits and patterns will have a more prominent impact on land use than population growth and will become the principal driver of land use change over the next few decades. Demand for meat products is rising rapidly in societies that have been largely deprived of them by poverty and poor distribution. Third, accelerating climatic changes and natural disasters is likely to result in more extreme weather patterns, with average temperatures and rainfall increasing, leading to fluctuating yields and production shortfalls. Fourth, there is significant waste along the supply chain with estimates as high as 30% globally, which, in turn, drives global demand on an already under-pressured agriculture supply chain.

In the animal supply chain, there is a growing focus on recording treatments particularly of antimicrobials, and in most developed world territories, there are mandatory requirements to record these treatments. As concerns about food safety grow in an engaged consumer society, the importance of proving provenance of supply has become important. It is a legal requirement to maintain a record of animal treatments that can be inspected by government officials and is a contractual requirement in many food supply chains. Auditors can come at short notice if they suspect a breach of the withholding times of antimicrobials has occurred. In the dairy supply chain, testing for the presence of antimicrobials in milk is routine, and contamination of a bulk sample can lead to expensive compensation payments to the milk buyer. A 20,000 L lorry tank of milk contaminated by a penicillin will cost a fine of over €6000 at current prices. As each antimicrobial treatment has different withholding times, good record keeping is essential and has to be maintained by farmworkers in a wet dirty environment. A common occurrence is for the notification of an audit to trigger a frenzy of searching through diaries for barely legible notes, backed up by invoices and prescriptions from the veterinary surgeon.

The food product supplied by the farm will pass through a number of owners and processors before it arrives in a shop to be purchased. The resulting agriculture value chain is extremely complicated and incorporates a wide array of stakeholders including, input companies, farmers, traders, food companies, and retailers, all of whom must ultimately satisfy the varying demands of the consumer in a sustainable manner. The sector encompasses huge diversity and variety at each stage, from R&D-based input companies to generic manufacturers, subsistence farmers to high tech agribusiness holdings with large staffs, biotech boutiques, and small and medium-sized enterprises (SMEs) to multinational corporations. The sustainable production of more food for human consumption requires technology that makes better application of limited resources, including land, water, and fertilizer. The essence of new technologies is that every artifact has digitized data attached to it. In animal agriculture, particularly many processes are conducted by humans in unstructured environments, such as in treating a sick animal, so creating the digital record of the event relies on fallible, forgetful humans rather than relentlessly consistent automation.

Technology adoption in agriculture is often cited as a real tool to assist the industry address the challenges it faces. This chapter offers an example of a way of thinking about marrying the use of several data capture technologies and stakeholders in the food supply chain and identifying reward mechanisms for the various stakeholders.

In this paper, we describe the system developed by the authors to improve the recording of animal disease treatments to meet both management and legislative requirements. We use this model to discuss how to manage the requirements of the stakeholders such as the ownership of data and those who add value to the data. We discuss who benefits and how a fair market system can be developed and protected.

Data—who owns and sells it

The concept that data have value that can be traded has been pre-empted by the "free service" providers of the Internet 2.0 model such as Google, Facebook, and a host of imitators who offer a useful service such as searching for information or connecting with friends and family. The free service is paid for by advertising, which is tuned to the users interests by analyzing their data.

The raw data in itself have limited value, and there is significant cost associated with collecting, aggregating, storing, and processing the data. There are legions of software developers, systems managers, standards setters, and a huge hardware network that is expected to be constantly available and constantly expanding on demand. All of this is paid for by the advertisers who have been handed a huge advantage over traditional advertising channels by getting feedback on the purchase decisions made after viewing the advertisement. The companies can also argue that the users can turn off the advertisements and personalized data processing. The argument is that the data are freely given by the users Fig. 10.2.

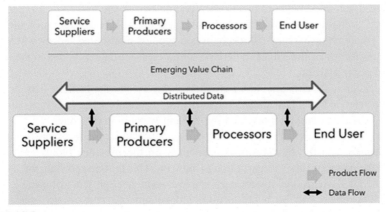

FIGURE 10.2

The emerging value chain now includes data, which link every process that occurs between field and consumption.

The loss of ownership, control, and rights associated with data both personal and business has been questioned by Lanier (2014). The data market and its components discussed in detail by the Economist, 2017 (Briefing, Fuel of the future Data is giving rise to a new economy How is it shaping up?). How to assign value to data for example in the balance sheets of companies is now a regular topic in The Economist (2020). The value of the average users' data has been calculated based on the advertising revenue divided by the number of active users.

The rules and acceptable practices for data ownership, control, and monetization are evolving for individuals, organizations, and governments. Perhaps, it is most useful to strive to adopt guidelines of how win—win use cases can be created in an agricultural setting. Copa Cogeca EU umbrella organization of farming representative unions and groups in Europe offers a comprehensive guide to the use of agricultural data.

Creating data transparency—blockchain methods

Blockchain is a "blossoming technology" as per FoodLogiQ. It is the technology powering Bitcoin and other virtual currencies. To understand blockchain, we will have to talk about the lesser known term distributed ledger technology (DLT). Although used interchangeably, DLT is the parent technology of Blockchain. Blockchain describes a) a type of distributed ledger—a tamper evident and tamper-resistant digital ledger implemented in a distributed fashion (i.e., without a central repository) and usually without a central authority (i.e., a bank, company, or government) (Yaga et al., 2019); although, not every distributed ledger is a blockchain.

Distributed ledger is the concept of having a distributed database as opposed to a traditional centralized one. The computers that the ledger is distributed across are called nodes. All these nodes spread across diverse locations, networks, and boundaries, contribute to the ledger. If there is a change made to a node, the ledger gets updated. Each node behaves independently, and hence, the update occurs independently. The main idea behind the distributed ledger is decentralization of authority, hence greater transparency. Blockchain is a type of distributed ledger where each node gets a copy of the ledger. Whenever a new transaction gets added to the ledger, all nodes get updated.

Blockchain has a distinct block structure, which is not the genuine data structure of distributed ledgers. Distributed ledger is simply a database spread across different nodes—the data can be represented in different ways in each ledger.

Each block in a blockchain is in direct sequence to its predecessor. In fact, the next sequence in the blockchain is dependent on the previous block and takes a part of the previous block's encryption to build on. This feature sets it distinctly apart from the DLT where data do not need to be in sequence, only distributed.

A blockchain can be a consortium blockchain ("a consortium created and controlled by a group of members"), a private blockchain (a "centrally controlled blockchain which only permits specific authorized members to add records to the transaction history"), or a public blockchain (the "original blockchain architecture where everyone can view the blockchain" and anyone can add/verify records and join or leave the blockchain) (FAO and ITU Bangkok, 2019).

The blockchain was created to record financial transactions in a distributed, decentralized, peer network to introduce transparency in the sector. Where money is involved, there is huge mistrust between the participating parties. "Thus, blockchain technology enables the technological institutionalization of values in environments that are highly dependent on these values". Three In terms of animal welfare, public health, food safety, and environmental health, agriculture is about as institutionalized as any sector can be. So, blockchain can be used effectively to gather, verify, record, and transfer information important to institutions whose values place protection of people, animals, and environment at their heart.

A consensus mechanism for when the blockchain gets updated with a new transaction was an immensely important requirement. There are different kinds of consensus mechanisms across the different blockchains. All of these consensus mechanisms consume huge amounts of processing power. DLT does not need such power hungry consensus mechanisms. All or some nodes can participate in a voting style verification of new information added to the distributed ledger, depending on the design and rules of the ledger. This makes DLT easily scalable, energy-, and budget-friendly.

For instance, in the AgTech industry, if we focus on the agri-food supply chain, DLT and then eventually blockchain have a huge role to play. It can help different players in the chain come together in an unprecedented transparency and collaboration of information.

Examples of blockchain in farming

So, far examples, of true blockchain, deployments in agri-food have been few and far between. The sector is not alone, as the announcement in May 2021 by Microsoft that they were migrating their Azure Blockchain to other Microsoft services. Here, we briefly outline two deployments of interest in agriculture. Carrefour4 organization is actively implementing blockchain solutions for its products traceability since 2018. The aim is to provide a QR code scan where by consumers can access data on the product with their mobile phone. The information available through the code includes place and date of production, the product's composition, method of cultivation, and more depending on the type of product. Carrefour expanded its blockchain provenance project to include textiles and has set a target that by 2030, 100% of natural raw materials used in "its exclusive brand products" will be sustainable and traceable. Carrefour also commits that by 2025, 50% of its cotton textiles will be organic. The retail giant has been working on blockchain with IBM Food Trust since 2018.

Starting as a Canadian beef platform, FoodLogiQ is now a full farm-to-fork system and enabler of Blockchain within its partners' supply chains. Using what it terms the "blossoming technology" that is blockchain, FoodLogiQ is open about the as yet glacial pace of uptake of blockchain within agri-food sector. They use hyperledger fabric as their chain's infrastructure, the open-source framework on which IBM's blockchain Platform is also built. FoodLogiQ places an emphasis on supplier management, food safety compliance, quality incident management, recall management, and whole chain traceability.

According to the FoodLogiQ website testimonials, Tyson Foods, a massive US-based meat processor, credits FoodLogiQ with helping them "test the applicability of blockchain within our supply chain." This reflects the state of play in the ecosystem at the moment; organizations considering the new technology and how it may be relevant and beneficial in their operations.

As illustrated in Table 10.1, these technologies have the potential to transform the agriculture value chain and provide the various stakeholders with significant benefits such as increased resource efficiency for farmers and greater involvement of the farmer as a proactive stakeholder in the value chain (Bronson & Knezevic, 2016), while at an environment level, these technologies can help to tackle issues such as

Table 10.1 Potential impacts for stakeholders of digital technology in the food supply chain.

Stakeholder	Positive impacts	Negative impacts
Farmer	• Increased productivity • Increased efficiency and cost saving • Supported decision making • Improved livestock health • Greater food/income security	• Data ownership • Inequality of value chain participation • Changing the expectation of what farming is • Loss of decision making power • Amplification of the digital divide
Processors	• Real time data and traceability • Better trust in products • Better quality products	• Exclusion of data literate producers • Increased dependency on technology
Retailers	• Increasing the transparency of supply chain	• Increasing market dominance in the value chain
Consumers	• Real time data and traceability • Better trust in products • Better quality products	• Changing food patterns • Voice or concerns not taken into account
Policy makers	• Greater standardization • Increased transparency across value chains through track and trace • Improve the efficiency and effectiveness of existing policies and programmes, and to design better ones • Move toward more targeted policies which pay (or penalise) • Enhanced monitoring compliance with animal welfare, environmental and public health policies	• Policy not keeping ahead of the fast pace of technology advancements • Digital technologies raise questions about privacy, interoperability, and even potential liability issues, all of which will need careful consideration. • Amplification of the digital divide

Table 10.1 Potential impacts for stakeholders of digital technology in the food supply chain.—*cont'd*

Stakeholder	Positive impacts	Negative impacts
Environment	• Enable automation of administrative processes for agriculture and the development of expanded government services • Increased efficiency of resources utilization • Reduced carbon footprint • Reduced food losses in the supply chain • Greater eco-efficiency	
Social	• Increasing the transparency of supply chain • Increased access to information to improve ability to respond to threats such as AMR or zoonotic diseases	• New market and business models that could lead to the lack of benefit sharing • Data ownership
Economic	• Increased productivity and profitability • Manage supply and price volatility • Empowering farmers in the value chain	• Increasing risk of market dominance by a few large multinationals

Source: Authors—Adapted from references.

natural resource utilization, reducing food losses in the supply chain, and in general greater eco-efficiency and sustainability such as reducing the carbon footprint of agriculture (Lynch et al., 2021). Broader access to farm-level information can be facilitated by these technologies, aiding public health responses, most immediately in the monitoring of antibiotic and antimicrobial usage at farm and indeed animal level, contributing to greater knowledge of potential risks and reservoirs of antibiotic resistant bacteria. From an economic perspective, Agri 4.0 technologies will enable greater utilization of data to optimize the supply chain for greater risk management analysis to respond to demand, supply variations, and price volatility in the market-place (Sonka, 2015). Nevertheless, there are potential negative impacts of Agri 4.0 technologies. For instance, it raises concerns over data ownership and privacy (refs) especially around legal rights and data access. For Wolfert et al. (2017), the emphasis on Agri 4.0 and big data could further move decision-making power from the farmers and other stakeholders into the hands of large private companies who have control over such data. Research has already highlighted that Agri 4.0 technologies are further amplifying this digital divide and inequitable benefit sharing across the value chain.

However, as Lynch et al. (2021) highlights the advance in digital technologies can also be a double-edged sword because the real danger is that new technology will further amplify the consolidation that has occurred in the twentieth century, whereby a small number of very large actors (Monsanto, Bayer, Cargill, Budge, Walmart, Tesco etc) dominate the value chain. While policy makers recognize that encapsulating agriculture data value chains requires a redesign of market models to enable a fully functioning and trustworthy data sharing ecosystems among stakeholders (EU Data Strategy), it may not be in the financial interest of large agriculture corporations who have gained so much market power and influence that they have the ability to shape markets and policies. Indeed, the list of the world's largest 500 companies contains a large number of agriculture and food firms who through mergers and partnerships are concentrating and exerting more control over every aspect of the value chain from farm to fork. Indeed, the agriculture food market could be reasonably described as an oligopoly.

The emergence of Agriculture 4.0 digital technology heralds the introduction of a multitude of new technologies and inevitably a host of new embryonic platforms to the agriculture marketplace. The introduction of these smart agriculture technologies and governance sharing models has the real potential to transform the historical agriculture value chain itself and to become a data-driven agriculture ecosystem in which all stakeholders can benefit through enhanced productivity and profitability.

This now raises perhaps the most fundamental question facing the agriculture sector—how will the agriculture value chain adapt and evolve in order to reap the fully benefits of Agri 4.0?

Agriculture 4.0 market models

The evolving Agri 4.0 market model can be equated to a technology and data battle ground whereby the various market stakeholders are vying for acquisition of the market share for farm data. With this view of the emerging Agri 4.0 market place, at least two business model scenarios emerge, which can be characterized as embryonic and mature market solutions (See Lynch & Quealy, 2017).

Model 1: closed business system

The mature market landscape is the most dominant business model of today and is characterized by a closed business system mentality where there is high customer engagement but low collaboration between actors in the business ecosystem. This scenario represents business models such as those presented by mechanized farm equipment manufacturers who progressively moved toward digitized, software-controlled components that require authorized software for access to repair, as well as restrictive contracts that forbid repairs and modifications. In this business model, the technology providers try to preserve their market position (e.g., prevent clients from switching to other providers) or to grow it (e.g., to get new clients). The focus of the companies in this marketplace is market share and protecting intellectual property via their own siloed technology platforms.

Predominantly, this market model is dominated by large technology and software companies who are creating market platforms consisting of intricate ecosystem of data flows across products, machines, farmers, and external partners and the technological infrastructure to facilitate them. On the whole, stakeholders have to pay a price to get access to use the product and the knowledge and are not allowed to interfere with the proprietary technology. The key business challenge with this value chain business model is that the agriculture industry will be characterized by knowledge silos and due to the lack of semantic interoperability that exists among different information systems, it will be difficult for stakeholders in the value chain to exchange the data in a meaningful way both at their local operating environment but also at a global level.

Model 2: participatory markets

Conversely, embryonic platforms such as VirtualVet are characterized by participatory market model, low customer engagement, and market share and the potential for high collaboration in the business ecosystem because these are often new and embryonic technologies that are been brought to market by start-ups and emerging platforms. The technologies entering the marketplace in this scenario are most likely to be ICT platforms used to manage data sharing, which is beginning to see the introduction of a number of aggregator type companies focusing on data for targeted markets across the different agriculture verticals such as health. The fundamental premise of this type of business model is the connected value chain based upon a participatory framework whereby the "value chain adopts a network concept where actors collaborate to co-create value that could not be created individually" (Lynch & Quealy, 2017).

In comparison to the traditional mature market model, the emergent platforms have the potential to dramatically reshape the value model of the industry and the data-driven business models required by this participatory market model. In this agriculture industry platform model, information will flow in multiple directions within the value chain resulting in an exponential increase in the quantity, frequency, and quality of data generated by consumers. Sophisticated new business platforms will emerge to support the capture and exchange Fig. 10.3.

The impact of government policies on supply chains

Will embryonic solutions survive? What can governments do to enable embryonic solutions survive and thrive?

Three key questions highlight the actions needed from governments to ensure the opportunities offered by digital technologies are realised:

- First, how can government policies and programs appropriately facilitate the adoption of digital technologies by the agriculture and food sectors? Policymakers will need to consider potential benefits, costs, and risks and to

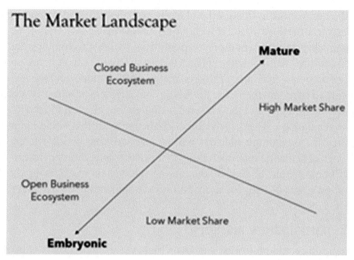

The Market Landscape

Mature

Closed Business
Ecosystem

High Market Share

Open Business
Ecosystem

Low Market Share

Embryonic

FIGURE 10.3

In mature markets, large business entities have high market shares and try to maintain a closed business ecosystem retaining data within their supply chain. Newer business models can disrupt this by using a more open business ecosystem.

understand the factors affecting technology uptake so that interventions can be targeted to where there is a market failure or a public interest. For example, verification of food labeling claims as outlined above.

- How can governments make use of digital technologies to design and deliver better agricultural policies? This requires understanding how technology can help in different components of the policy cycle and may require government bodies to expand their skillsets, invest in technology and training, or partner with other actors (both government and nongovernment)
- How might digital technologies change the roles of government? On the one hand, digital technologies may create new roles or responsibilities for governments, including to enable the digital infrastructure (is there a case for governments to be a provider or a rule maker of new digital infrastructure, and under what circumstances), but on the other hand, if technology can reduce information asymmetries and transactions costs, less government intervention may be needed. A functioning and participatory data market could contribute to more transparent supply chains and pricing models, which reward openness, fairness, and quality from farm to fork.

For policymakers, the challenge will be to shape policy and regulatory settings so that they facilitate opportunities offered by digital technologies. At the same time, and not unique to the agriculture sector, digital technologies raise questions about privacy, interoperability, and even potential liability issues, all of which will need careful consideration.

In 2021, the OECD issued a report on government supports to agriculture, which suggested considerable room for improvement in the approaches commonly taken. The report finds that "in some developed countries, support remains high and linked to production, while some emerging economies have also significantly increased policy interventions that distort production decisions." In both cases, support could have been better targeted at public services that benefit producers, consumers and society overall. The opportunity now if for government to be a customer of innovative solutions, which can be deployed at scale as public services. These targeted innovative solutions deployed in this way can contribute to States' need to address challenges outlined earlier. Embryonic solutions and data platforms have most likely already benefited from state funded research or start-up support, partnering with these companies to increase the effectiveness of responses and grow employment.

There is a bewildering range of claims and standards emerging in food and protein systems. Several of these are driven by large, institutional investor groups, for example, the Taskforce1 for Climate Related Financial Disclosures (TFCD, 2021), recently signed up to by the UK government. Food retailers, from supermarkets to restaurant and food service companies, are coming under increasing scrutiny with pressure mounting on them to display detailed knowledge of their supply chain's risk profile on certain topics—animal welfare, antimicrobial resistance, greenhouse gas emissions, water protection, etc.

In parallel, the European Commission and governments across Europe are attempting to demonstrate national or State-level verification of policy implementation on the challenges mentioned above. This sounds like the ideal opportunity for innovative start-ups and solutions to prosper, serving, and partnering with busy incumbents to meet the needs of incredibly important areas: food production, animal welfare, and public health.

However, when governments create reporting structures for the capture of digital information, it has two main negative impacts on innovation: legislative clout to override private market services providers and the creation of a national silo with neither the remit nor habit of putting data to commercial use.

European union veterinary data market

Taking as an example the monitoring and surveillance of veterinary medicine usage, in particular antibiotics. The EU regulations on veterinary medicines 2019/6 will come in to force across member states from January 28th, 2022. There are currently very few standardized or agreed methods of collecting veterinary medicine usage in food producing animals. The new EU regulation is in response to calls from the WHO, UN, OECD, and others calling for much improved antibiotic and antimicrobial stewardship along the food chain. AMR, the ability of infections, bacterial, fungal, and viral, to survive and thrive in the presence of treatments that were once effective at killing them, kills tens of thousands of people worldwide each year.

In Germany, a study of eight years of antimicrobial usage in veterinary medicine came to the conclusion2 that there was a gap in true usage information, but a system was in the pipeline as a response to EU Reg 2019/6. In Ireland, the Department for Agriculture, Food and the Marine announced in late 2020 that it would create a National Veterinary Prescription System to capture all veterinary prescriptions. However, the intention is to a) only record a maximum of four tag numbers (animal IDs) per prescription and b) there is no intention to make any specific trends, patterns, insights, and available to the wider agri-food industry. A new silo is built, overlooking the multiple farm level and vet level digital services already working in the space. These small companies cannot compete with the state.

Exemplar: the livestock production chain

There is insufficient space to describe every different production system, but we can take as an example the traditional systems of grazed livestock production that have changed little during the agricultural revolutions in Northern Europe. A typical farm in the beef sector in UK and Ireland will have about 50 breeding cows. Every year, the cows will produce a calf, which is suckled on its mother for a few months before it is weaned and fattening begins. Most farms still use natural breeding with a bull, and so even insemination dates will not be recorded. Animals live a substantial time in open fields with little more supervision than a daily count. This is the type of production that has genuine quality credentials in the eye of the consumer, sequestering carbon and having high animal welfare. Sometimes, the calves are sold to feeders on another farm. By law, the calves have to be tagged with a nationally issued identification number at birth, and every movement between farms has to be recorded to be able to trace disease vectors for control programs such as those to combat tuberculosis and other infectious diseases. The recording system has to support a legacy paper-based book system with documents sent by post for those who do not use a digital connection, and this is slowly transitioning to all being stored digitally. Many farms may not have access to the Internet either through lack of connection or willingness to use computers. The average age of the farmers is high, which means they have not had exposure to the benefits of computers until later in life. The security of the data in the national database of movements is a responsibility of the government who absorb the maintenance cost. There is a natural hostility of the independent minded beef farmers to government and buyers monitoring their behavior and possibly punishing them for accidental infringements that can happen under the pressure of managing livestock in a dynamic environment. Ultimately what we end up with is a closed and disjointed value chain where all the stakeholders are working independently of each other and competing among themselves for the farmer's data, attention, and revenue models. This lack of trust, cooperation, and collaboration means that the farmer's resources are treated like a transaction to be won and not shared with others resulting in an inefficient system for all, especially the farmer. So how can we improve this situation for the farmer and the processors who need uncontaminated raw materials to which to add value? Fig. 10.4.

FIGURE 10.4

Data which are generated by the farm and its suppliers can be attached to a product and has a value to processors, markets, governments, and in bulk to pharmaceutical suppliers.

Virtual vet system for recording treatments

The widespread adoption of data connected smartphones after 2010 led to a demand for apps. The VirtualVet idea in 2015 was an app that could be used to record veterinary treatments on farms next to an animal when it was being treated. Data would be sent immediately to our database so that farms could easily print out records of which animals had been treated and when. We knew enough about poor farm connectivity such that the app only needed intermittent connection to the internet (for example, in the coffee shop where the vet stopped after his morning round); however, a number of issues arose.

The names of pharmaceuticals are long and hard to type, and the range in use far exceeds the length of a sensible drop down menu system. Animal work often requires two hands, and so the data were still being collected after the event. The light conditions are either too sunny or semi dark for easy reading of screens. The work often means that gloves are being worn, or it is too wet from rain or milking parlor wash down for operators to risk their expensive personal phones. The feature that worked best was the image capture so that a farmer could photograph the ear tag and the label of the medicine in use.

Any data entry gave us a time and user, and from this, we could deduce the herd data, and we evolved to curating data with a call back and making sure that data entered onto the main database met a good quality standard. The arrival of messaging apps enabled us to abandon support for the VirtualVet app, which also became expensive to update with ever changing OS systems.

The process evolved into the farmer sending messages with images of ear tags and medicine labels or audio messages full of the information they knew we needed. We needed human curators to answer the phone and transcribe the data, but this did enable us to be confident that the reports we produced were well structured, and every reported incident was consistent to a quality standard (the medicine spelled correctly and treatment reason spelled in searchable format).

We were able to start linking data for example to the national database of animal IDs, which also enabled a qualitative improvement since we could then spot IDs, which were not consistent. The call back also allowed us to record the effect of the treatment (recovered, partly recovered, died, etc), which is very rarely the case. This allows us to collect data on the effectiveness of treatments.

We were able to find some clients in the veterinary profession and supply chain who valued the data reports, which satisfied their need to manage the data flow around veterinary pharmaceuticals. The farmers who had done the work of sending us the data benefited by being able to print out audit reports without the scramble to find the diary entries. They were also able to trace the animals that needed the most treatment which allowed a Pareto analysis of animals being treated Fig. 10.5.

While the human curation process is hard to scale, we believe that new technology (OCR and voice analysis) will overcome this limitation in the future, and our skilled staff will be more focused on training machine learning algorithms to analyze messages received from the farm.

Farmers are price takers and have surprisingly little control on the economic viability within their farm gate. Margins on beef animals are low, as entry is easy,

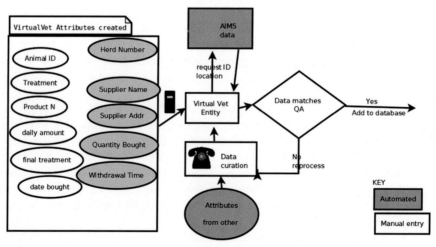

FIGURE 10.5

A simplified entity and attribute relationship diagram for the VirtualVet system. There is minimal data needed from the farm or vet, much can be deduced or verified against other data such as AIMS, which records animal IDs and movements in Ireland.

and there is plentiful supply of frozen beef from distant sources. So the question of who pays for the time taken to record data and who pays for the data aggregation process that VirtualVet does. Developing a sustainable pricing for all stakeholders is a key issue. What additional value does a data tag add to piece of beef or a liter of milk? This is a question that is still unanswered, but we believe that the consumer has to bear some of the extra cost of provenance data.

Agriculture 4.0 technologies in the value chain

Digital technologies are penetrating the agriculture market at an accelerated pace. It is estimated that there are approximately over 75 million agriculture IoT devices currently in use, and it is expected to grow to over 225 million by 2024 (Andrew et al., 2018). While there are numerous technologies, there is a general consensus in the literature and most practice, the four main disruptive technologies are:

- Internet of Things (IoT): Refers to the use of a family of technologies such as sensors, control systems, motion detectors, cameras, and many other devices that converts actions involved in agriculture production and value chains into large volumes of data, whereby real-time insights about farm and value chain productivity can be made available. Indeed, the aim of most agriculture IoT products is focused on enabling the farmer to use data insights to improve control and increase the business efficiency by making better operational decisions. For instance, sensors and drones could be used for crop and herd monitoring or providing insight into which plants need spraying to improve their health and which do not.
- Automation and robots: The field of automation and robotics in agriculture known as "agribots" has garnered significant interest because the automation of specific operations such as milking, planting, inspection, etc., enables farmers and other stakeholders to overcomes issues such as the aging farmer population, the shortage of agricultural workers, and the rising cost of labor. The focus of agriculture automation has been primarily on autonomous vehicles such as tractors and in the main are designed to remove manual intervention in agriculture to enhance the efficiency of the agricultural operation.
- Artificial Intelligence (AI): Within agriculture, AI is an emerging technology, which is software that shows intelligent behavior by combining problem-solving and decision-making to achieve specific goals. It is developed to be used in a range of applications including satellite imaging, animal health, heating and feeding machinery, etc. Through machine learning, the AI system learning is enhanced by the machine's exposure to new scenarios and data, which, in turn, increases its decision-making capabilities without being explicitly programmed.
- (Platforms) traceability and blockchain: A blockchain is a method of ledgering in a distributed fashion. In essence, it is a chain of information blocks interconnected in a digital environment allowing for information on transactions of various kinds to be stored, linked, and recovered, forming a large database. The

escalating challenges facing the agriculture industry has accelerated the use of block chain in the agriculture value chain because it is seen as a mechanism for optimizing different issues, such as transparency, cost effectiveness, traceability, quality supply systems, etc. Because blockchain is a platform-based technology, it has the potential to promote more sustainable agriculture supply chains, better facilitate trade, and provide a more inclusive trading and new market opportunities for micro-, small-, and medium-sized enterprises.

Agri tech vendors, nodes in the embryonic participatory model

VirtualVet was developed from 2014 onwards as a data collection, digitization, management, and visualization service for the agri-food industry focused on veterinary medicine information. The business model is novel in agriculture; it is a free service to farmers, paid for by those using the information along the value chain. Notable successes and learnings with application to several sectors within agriculture. In the context of this chapter, VirtualVet is a node or ledger. The platform can cooperate with other services where there is a willingness to meet the needs of the farmer, industry, and wider public health. The participatory market model enables the capture and exchange of data. VirtualVet learned early that presenting farmers with an app and expecting full and total accuracy and timeliness is both selfish and unrealistic. With the extent of regulatory responsibilities and compliance pressures hurtling toward farmers at an ever increasing pace, loading another task on to farmers is not really helping. We instead looked at farmers' habits and met them where they are technically and time-wise. So VirtualVet clients can text, call, leave a voice memo or request their vet email us a copy of their prescription. This flexibility and often "manual" intervention to interpret and curate analogue information in advance of digitization is an early stage in the digital transformation path we describe below. It is no less important just because it is bespoke. With commitment from stakeholders, real scale is achievable, again following the digital transformation process set out below.

How to improve supply chain provenance

Blockchain is being praised as a technological innovation, which allows to revolutionize how society trades and interacts. This reputation is in particular attributable to its properties of allowing mutually mistrusting entities to exchange financial value and interact without relying on a trusted third party (Wüst & Gervais, 2018, pp. 45–54). With blockchain, no data will ever be lost. Once an action has been performed—an animal sold, an animal treated, sent to the factory, it gets updated on the ledger and stays. Even DLT without the full blockchain experience would be a considerable leap forward in the collaborative and cooperative journeys, which must be taken by large, dominant business players in their closed systems. The examples presented of blockchain active in some supply chains, representing a small portion of each partner's overall supply chain are excellent places to start, but more progress needs to be made.

Taking the veterinary medicine regulation (EU Regulation 2019/6) as an example. The EU will require member states at first, to report veterinary prescription information. This will evolve to actual usage information in a couple of years. The regulation is intended to address some information asymmetries in the sales, supply, prescribing, and dispensing of these important drugs. For example, there is currently no easy means to report the poor performance of a veterinary medicine. In cases of extreme dissatisfaction by a farmer or vet, perhaps the manufacturer of the veterinary medication will send a representative to meet the vet and/or farmer and discuss the drug's efficacy—or lack thereof. Neither the farmer nor the vet has any way of knowing for certain how many similar cases of dissatisfaction are currently being dealt with by the manufacturer. Was there a bad batch? Was the product tampered with? Is the product simply not fit for purpose in the real-world despite success in limited trials. A farmer and vet focused collaboration of actors in the animal health chain can highlight and make available to those permissioned to access, a wider pool of insights about a veterinary medicine, it is effectiveness and any contraindications.

For a wider public health demonstration, imagine if researchers and hospital laboratories could access see, on a near-real time basis, the geographic spread of antibiotics being administered to animals, including the reason for that usage. Imagine if hospital laboratories could also access the sensitivity testing results carried out by veterinary laboratories in advance of issuing antibiotic prescriptions, thereby giving early notice of the presence of a bacterial infection in an area and a list of the antibiotics to which the bacteria has developed resistance. The OneHealth approach considers that human, animal, and environmental health are intrinsically linked and impact on one another; for better or worse. AMR is a global threat to human, animal, and environmental health.

Without the ability to directly query, discover, and assess the validity of data and information in any system, it cannot be fully trusted. Closed systems such as exist in the agri-food and animal health industries at present are not best placed to respond to not plan mitigation strategies against the challenges facing the industry and its stakeholders: all of us.

Anything less than an accessible platform runs the risk of being written off as a public relations exercise.

An embryonic blockchain platform—veterinary medicine surveillance

The creation of an information exchange platform is required. In this example, we set out how a DLT-enabled platform could be created, offering existing AgTech companies, agri-food processors, animal health companies, and government agencies the opportunity to meaningfully collaborate on animal welfare and AMR. This requires a patient but purposeful movement from a starting point in the "Current State" enabled by manual or paper-based agreements and email or .CSV files exchanged as and when agreed to a "Target State" of automated and harmonious information exchange. The five issues presented in the Current State are that data, information, and indeed processes are analogue, manual, inefficient, distributed, nonnormative, and siloed. Each of these sates of information and processes can slow down at best and at worst introduce inaccuracies in to an operation. To be clear, while "distributed" here is seen as a negative, it is because there is no agreed process, standard, or format in these distributed databases or information stores. They present only part of a picture. The move to centralized in this scenario is not as a single centralized repository; it is more of an agreed bringing together of information for an agreed purpose; a complete picture.

The five steps we suggest walk stakeholders from assessing their current documents (physical and electronic), emails, needs, deadlines, and common practices to identify how the agreed objective can be met by a more digital process. From this foundation stage, the stakeholders move to proving the value of actions such as replacing filing cabinets of documents with databases of accurate information captured from web services. Next, we introduce the buyers. Here, buyers are really customers, both internal and external to an organization. They may purchase access to the information with money or by exchanging a service.

It is at this stage that stakeholders start to get a real glimpse of the potential benefits for their operations. Now more organizations want to be included, and scaling takes places. With scale comes economic certainty to invest and develop more and more automation of the processes, achieving the target state. In this target state, data and information are digital, automated, optimized (from the best sources), centralized, standardized, and harmonized.

Such a digital transformation would create a platform for data exchange, with each data source along a value chain now acting as a ledger or node in accordance with the agreed and standardized processes relevant to that value chain. In the animal health sphere, this would enable farmers and those who serve them or buy from them to act as data sources and information recipients as and when appropriate. The standardized digital information can now be accessed by buyers; those organizations with a need to know, such as food processors, retailers, government agencies, or public health researchers.

Such a digital transformation would enable AgTech vendors to specialize in their area of expertise, delivering their products and services to farmers, while facilitating, for a fee, access to agreed answers from others in the value chain. An example marketplace in veterinary medicine data is illustrated here. The sellers of information cooperate with a platform to present standardized and complete records and insights. They offer access for sale to buyers with a need to answer questions consistently and robustly. The revenue, at scale, can be shared on an agreed basis.

Discussion and conclusions

Agriculture is changing. It is responding to the scientific and policy demands to reduce harm caused to the planet by the production of food in identified systems. The requirements to meet targets and comply with legislations and consumer expectations are increasingly continuously. AgTech vendors must recognize the human toll this takes on farmers, growers, and producers as they navigate a changing sector; exhaustion, stress, and economic penalties to name a few. Such is the extent of this toll that the EU lost 500,000 farms between 2015 and 2016, with the annual Agriculture, forestry, and fishery statistics (2020 edition) commenting on the steep decline in farm numbers "for many years." It is an untenable business model to sell to such a staggeringly and rapidly reducing number of buyers in the ing market. It is time for AgTech vendors to collaborate, creating new value from their existing products and services by trading access to data on behalf of the farmer, and ensuring some economic value is also shared to the data source; the farmer.

A lingering source of frustration for those seeking greater transparency in food systems is the continuing resistance among AgTech providers and creators to producing a truly farmer centric offering. Data are collected, stored, and used for the purposes of their device or product. The willingness to act on behalf of a farmer is missing. Instead, vendors convince themselves that by presenting farm level information back to a farmer meets their requirements under a simple transactional relationship.

This single user—vendor relationship is not fit for purpose now, and certainly is not fit for the rapidly approaching future of verifiable traceability and provenance claims. Unless AgTech providers actively invest in placing the farmer's needs at the center of their offering, they will run the risk of becoming an inconvenience to their farmer customers. Or worse, irrelevant. They also miss the opportunity and responsibility to be an active participant in the drive for a better food system.

Presenting information to a farmer for the farmer to then manage, access, and re-present to several compliance or regulatory bodies is no longer sufficient. What is needed a platform approach, where AgTech providers cooperate, combining the information from several of a farmer's other agri-tech providers. The service, which will request permission from the farmer to present the required answers directly to a compliance authority, will win in the eyes of a farmer already busy and reluctant to become a mere go-between in their own operation. The model has to be something similar to ISObus in farm machinery, which allows interoperability of different proprietary machines.

Instead a farmer can and should be a director, a conductor; managing and overseeing the performance of those he or she pays to ensure they are contributing maximum value to the farm business. Taking the conductor view further, where, say in an orchestra, players or sections do not work together, the results can be just noise. But when they pay attention to each other, understanding where one can stop and another take over, magic happens. AgTech providers need to know their score. They need to be aware of the compliance requirements their farmer clients must meet. With knowledge of the questions a farmer needs to answer, not just the questions a piece of AgTech can answer, AgTech providers can move from being merely a nice-to-have to a must-have.

In this view, the AgTech vendor as a trusted specialist and data partner with a farmer, grower, and food producer, can become a trusted node in relevant blockchains or distributed ledger technology—based platforms a farmer's produce moves along.

Blockchain and its underlying technology DLT are merely examples of existing tools, which can be leveraged by a cooperative AgTech community, acting within a platform to meet agreed objectives for a specified value chain. Agriculture 4.0 is here. It is exciting. And, it might just save lives and livelihoods.

AgTech and the wider agri-food industry can harness DLT and the philosophy of information exchange to enable as many actors in their sphere to meet the challenges and expectations facing the entire industry. It is time for real collaboration with a true and shared sense of purpose; to defend, promote, and improve our food production systems in a way that protects humans, animals, and the environment for as long as is possible.

References

Agriculture, forestry and fishery statistics 2020 edition, Eurostat, https://ec.europa.eu/eurostat/documents/3217494/12069644/KS-FK-20-001-EN-N.pdf/a7439b01-671b-80ce-85e4-4d803c44340a. Accessed October 14th 2021.

Andrew, R. C., Malekian, R., & Bogatinoska, D. C. (2018). *IoT solutions for precision agriculture, 2018 41st international convention on information and communication technology* (pp. 0345–0349). Electronics and Microelectronics (MIPRO). https://doi.org/10.23919/MIPRO.2018.8400066

Bronson, K., & Knezevic, I. (2016). *Big Data in food and agriculture* (Vol. 3). Big Data & Society. https://doi.org/10.1177/2053951716648174

Copa Cogeca EU. Copa Cogeca EU Code of conduct on agricultural data sharing by contractual agreement. https://copa-cogeca.eu/

Dicks, L. V., Rose, D. C., Ang, F., Aston, S., Birch, A. N. E, Boatman, N., Bowles, L., Chadwick, D., Dinsdale, A., Durham, S., Elliott, J., Firbank, L., Humphreys, S., Jarvis, P., Jones, D., Kindred, D., Knight, S. M., Lee, M. R. F., Leifert, C., Lobley, M., Matthews, K., Midmer, A., Moore, M., Morris, C., Mortimer, S., Murray, T. C., Norman, K., Ramsden, S., Roberts, D., Smith, L. G., Soffe, R., Stoate, C., Taylor, B., Tinker, D., Topliff, M., Wallace, J., Williams, P., Wilson, P., Winter, M., & Sutherland, W. J. (2019). What agricultural practices are most likely to deliver "sustainable intensification" in the UK? *Food Energy Secur, 8,* e00148. https://doi.org/10.1002/fes3.148. (Dutia, 2014).AgTech EU Regulation 2019/6.

Economist. (May 6, 2017). *edition, Fuel of the future, Data is giving rise to a new economy.*

Economist. (February 27, 2020). *The information economy, Rethinking how we value data* (FAO, 2017).

Lanier, J. (2014). *Who owns the future?* Publisher: Simon & Schuster (March 4, 2014). ISBN13: 9781451654974.

Lynch, P., & Quealy, S. (2017). Finding the value in an animal health data economy: A participatory market model approach. *Frontiers in Veterinary Science, 4.* https://doi.org/10.3389/fvets.2017.00145

(Lynch et al 2021) https://www.frontiersin.org/article/10.3389/fvets.2017.00145

The publication The future of food and agriculture − Alternative pathways to 2050 available at: www.fao.org/3/I8429EN/i8429en.pdf/www.fao.org/publications/fofa

https://www.ledgerinsights.com/carrefour-expands-blockchain-traceability-to-textile-products/

https://blogs.gartner.com/avivah-litan/2021/05/24/microsoft-ends-azure-blockchain-service-where-is-enterprise-blockchain-heading/

https://azure.microsoft.com/en-us/updates/action-required-migrate-your-azure-blockchain-service-data-by-10-september-2021/

Testimonial on FoodLogiQ's website https://www.foodlogiq.com/customers/. accessed October 13th 2021.

https://www.gov.uk/government/news/chancellor-sets-out-ambition-for-future-of-uk-financial-services

pone https://doi.org/10.1371/journal.pone.0237459

Rizvi, S., Pagnutti, C., Fraser, E., Bauch, C. T., & Anand, M. (2018). Global land use implications of dietary trends. *PLoS ONE, 13*(8), e0200781. https://doi.org/10.1371/journal.pone.0200781

Sonka, S. (2015). Big data: From hype to agricultural tool. *Farm Policy Journal, 12*, 1—9.

TFCD. (2021). *Web site states that The Climate Disclosure Standards Board is part of CDP Worldwide, registered charity number 1122330, a company limited by guarantee 05013650 and its wholly owned subsidiary CDP operations Ltd company registration number 06602534, headquartered at CDP Worldwide, 71 Queen Victoria Street, London EC4V 4AY, United Kingdom. CDP Worldwide is regulated by the Charity Commission.*

Wolfert, S., Ge, L., Verdouw, C., & Bogaardt, M.-J. (2017). Big data in smart farming — a review. *Agric Syst, 153*, 69—80.

Wüst, K., & Gervais, A. (2018). Do you need a blockchain?. *2018 crypto valley conference on blockchain technology.* (CVCBT). https://doi.org/10.1109/CVCBT.2018.00011

Yaga, D., Mell, P., Roby, N., & Scarfone, K. (2019). *Blockchain technology overview.*

Further reading

Amos, N., & Sullivan, R. (Eds.). (2017). *The business of farm animal welfare* (1st ed., p. 249). Routledge. https://doi.org/10.9774/gleaf.9781351270045

Fróna, D., Szenderák, J., Harangi-Rakos, & Monika. (2019). The challenge of feeding the world. *Sustainability, 11*, 5816. https://doi.org/10.3390/su11205816 (FAO and ITU Bangkok, 2019).

Meijer, D., & Ubacht, J. (2018). The governance of blockchain systems from an institutional perspective, a matter of trust or control?. In *Proceedings of the 19th annual international conference on digital government research: Governance in the data age (dg.o '18)* (pp. 1—9). New York, NY, USA: Association for Computing Machinery. https://doi.org/10.1145/3209281.3209321. Article 90.

OECD. (2021). *Agricultural policy monitoring and evaluation 2021: Addressing the challenges facing food systems.* Paris: OECD Publishing. https://doi.org/10.1787/2d810e01-en. . (Accessed 13 October 2021). https://www.oecd.org/agriculture/topics/agricultural-policy-monitoring-and-evaluation/

Zambon, I., Cecchini, M., Egidi, G., Saporito, M., & Colantoni, Andrea. (2019). *Revolution 4.0: Industry vs. Agriculture in a future development for SMEs. Processes. 7. 36.* https://doi.org/10.3390/pr7010036

Zhai, R., Tao, F., Lall, U., Fu, B., Elliott, J., & Jägermeyr, J. (2020). Larger drought and flood hazards and adverse impacts on population and economic productivity under 2.0 than 1.5°C warming. *Earth's Future, 8*, 7. https://doi.org/10.1029/2019EF001398. e2019EF001398.

Risk assessment of introducing Digital-Agritech

11

Toby Mottram

Digital Agritech Ltd, Kirkcaldy, United Kingdom

Surely, nothing can be more plain or even more trite common sense than the proposition that innovation [.] is at the center of practically all the phenomena, difficulties, and problems of economic life in capitalist society.

Schumpeter (1939), 87.

Introduction

It would be naive to assume that we can introduce a huge change in the operation and control of agricultural machinery, of animal production systems and food supply chain management without creating new and possible unforeseen risks. This chapter attempts to identify some of the risks and the potential mitigations.

The need to increase agricultural production to feed a growing population has been a feature of human development since prehistoric times. Throughout history, new technologies have come along and changed the ability of humans to produce food. There was a profound shift from hunter gathering to animal keeping and a further shift to cereal and horticulture production. Doubtless, there was some discussion about how these changes were implemented, but we have no direct records from pre-history. When farm machinery was first introduced in the United Kingdom, there were sporadic outbreaks of Luddism or machine breaking. More subtle risks to the environment such as the effect on pollution and ecology of 20th century innovations such as nitrogen fertiliser and insecticides were missed at their introduction and have led to much regulation and expensive clean-up operations. We need to avoid these when introducing digital agri-technology on a wide scale.

The risks can be broken down into introductory risks, operational risks and ethical risks. Ethical risks might be the change of spiritual relationships between humans and animals or between humans and the countryside and lie outside the scope of this engineering discussion. Similar ethical dilemmas must have occurred to farmers replacing horses with tractors during the twentieth century.

Girdziute (2012) reviewed and analysed risk of introducing agri-technology, largely focused on avoiding financial losses due to crop failure or animal disease or over-supply causing price collapses. In this chapter, the subject is risks inherent in the introduction and operation of new technology, particularly digital and robotic systems.

Aims of digital agri-technology

Innovation in digital agri-technology is driven by a number of factors created by our current technical and social situation

- the shortage of labour willing to work on farms,
- a desire to reduce chemical interventions in crops,
- reduction of greenhouse gas emissions,
- improvement of soil status
  - less compaction
  - more carbon and nitrogen sequestration
- supply chain standardisation and provenance by sensing
- improvements to animal welfare by removing the variability of human management
- improved farm margins
- a desire to build new companies with new technologies
- improved safety for operators and farm staff

At the end of the 19th century, the global population was estimated at 1.6bn. Crop production had increased dramatically by the extension of modern arable farming into the lands of hunter gatherers particularly in the Americas, Africa and Australia. The soil in those grassland and forested landscapes had built up reserves of carbon and nitrogen which when ploughed provided good yields initially. However, repeated cropping was leading to yields declining and there was an urgent need to develop alternative methods of producing fertiliser and to breed plants that more efficiently partition nutrients to grains. In the Wheat Problem (1917) first presented at the British Association meeting in 1898, Crookes discussed the declining yields of cereals grown on the newly cleared lands of Canada, the United States of America and Australia. The initial soil stores of nitrogen, phosphorus and potassium needed supplementation and the reserves of nitrate from guano manure from Pacific islands were almost depleted.

The invention of the Haber—Bosch process to synthesis nitrogen fertiliser, the application of the internal combustion engine and later the biological driven 'Green Revolution' enabled the global urbanising population to grow and thrive, so that now we have 7 billion humans with at the time of writing, stable food prices indicating sufficient supply. Powered machinery also developed but largely to replace the diminishing supply of labour. The people most at risk from the application of power to agriculture have been farmworkers and farming families. Any new developments in digital control of agricultural machines must be made safer than the existing machines which caused the majority of deaths and serious injuries on farms in the United Kingdom (HSE, 2021). In the parallel developments of autonomous vehicles, it is clear that the new systems may be held to a greater safety standard than human-controlled vehicles which are responsible for a huge number of deaths and injuries worldwide on the road network (Lee & Hess, 2020).

Increased food production to meet the needs of a growing population has come at a huge cost to the environment of the planet, to biosecurity and animal welfare and these should be conditions that digital technology can address.

- Monocultures driving out species diversity
- Pesticides causing insect and thus biodiverse populations to be stressed
- The release of methane from enteric fermentation and nitrous oxide from soil as a major contribution to greenhouse gas emissions
- Reduction in organic matter in the soil leading to less water retention
- Ground and river water pollution by N compounds and pesticide residues
- High energy consumption for cultivations
- High transport costs for fresh vegetables
- Reduction of animal welfare as animals were moved onto concrete and under cover
- Antimicrobial resistance through treating farm animals with antibiotics

The requirements for Digital Agri-tech engineering are that it must undo some of the damage done by the innovations of chemists and biologists whilst avoiding new problems which this essay hopes to explore.

Because problems emerge only after the introduction of a technology, I have had to use some imagination in identifying possible problems. My own experience in developing robotic milking has informed my thinking here as we created a risk sheet for what would happen if cows were largely milked by robots. Fortunately, none of the risks we identified caused major issues and the ethical fear that cows milked by robots would become wilder in the absence of human contact did not arise. In fact cows milked by robots are notoriously docile and do not run away from humans. We had misinterpreted the compliance of cows to human control as being based on friendship rather than fear of blows and shouts. There are doubtless similar assumptions that will prove to be unfounded as new technology takes over.

It is easy to think of engineering systems as being fairly benign such as replacing chemical weed control with automated hoes or zapping them with lasers, microwaves or plain old electricity. There are also external risks of perception by the politicians and lobby groups that can have a serious impact on the legislative framework and marketability of agricultural products.

Failures in risk management in genetic agri-technology

An unanticipated risk of introducing some technologies was the opposition created by pressure groups and politicians and the industry needs to be careful how it presents the new technology. Farm animal cloning and genetically modified crops have had difficulties becoming widespread because of legislative blockages. The first successful clone of an animal was Dolly the sheep at the Roslin Institute

(Mummery et al., 2021, pp. 131–158). It was largely curiosity-driven research to develop techniques that could also be used in genetic manipulation. This technology is now 25 years old and it is still not main stream with only about 6000 cloned animals worldwide, mostly used in research. Risk assessments indicate that food from cloned animals is no different from that of naturally bred animals. There are multiple factors that have inhibited the uptake of this technology including technical issues that are not fully understood and therefore the clones may not realise their potential. The cloning work at Roslin did not lead to great IP opportunities for the institute which is now absorbed into the University of Edinburgh and does not mention cloning in its research profile. There are a number of biotechnical and ethical issues in cloning which are yet to be overcome.

Cloning possibly created a 'yuck' factor a social response of distaste from the wider community. The idea of fields full of animals with identical genetics just did not accord with human perception of what was right. It was interfering with nature, playing God, etc. Those attitudes may change in future, but we should definitely be aware that there is a high risk of societal rejection for new technology in food production resulting in regulations.

Newspaper and political campaigns against 'Frankenfoods' led to the banning of genetically modified crops in Europe and some other territories despite their potential advantages in enhanced micro nutrients, disease resistance and increased yield. Developers of digital agri-technology need to be aware of this potential resistance and ensure that they explain and understand the risks of the new technology. Some of the risks may not directly affect consumers or users but may be interpreted by powerful groups in society as harmful and politics used to prevent introductions by legislation. There is a growing acceptance of the concept that just because a technology is possible it should be assessed for its impact before introduction. The challenge for the authorities is to keep up with technology before it is introduced; this is all the more difficult with the invisible nature of software. Only an expert can tell whether a machine is controlled by software or the human sitting in the cab.

The introduction of machine learning or artificial intelligence in many applications is a case in point where it has been shown that inherent bias in algorithms may disadvantage applicants of some ethnic backgrounds (Akter et al., 2021).

The risks can be broadly broken down into categories of risk:

- design risks — will it work as planned
- operational risks — can it be reliable in farm conditions
- ethical/legislative risks

Design risks

A frequent response from engineers when asked if something is possible is to say 'nothing is impossible in engineering if you have enough money and time'. However, the commercial designer has to consider the price that a final product can be retailed at so that farmers can enjoy a margin after purchasing it. Very few products in agricultural engineering are made at sufficient scales, especially in the early intro-duction phase, to be as cheap as consumer products. Traditionally, we rely on inno-vations in other sectors to make components cheap enough to build cost-effective systems, for example, motors, microprocessors, computer Operating Systems, cam-eras, sensors, hydraulic and electric drives are all available off the shelf. Prototypes can be put together relatively cheaply; the main difference tends to be the sensing requirements for plants and animals and the software to analyse and make decisions; these are inherently application specific. Software tends to need to implement mech-anistic models which have been derived from agricultural research. A simple example say of a planting system might expect the matrix of soil moisture, soil nutrient status, crop growth rates, weather prediction, seed requirements, etc., to determine when to drill a crop. When the parameters are right, the system would switch itself on and go and plant the crop. Such a mechanistic system could work well but farm experience indicates that other factors outside the known parameters pertain (e.g., seed not delivered on time) and the machine might just never leave the shed if the full mechanistic set of criteria do not match the model. Farmers often have to take risky decisions to get a crop into the ground and these may include rare unpredictable events. Or it might work perfectly 1 year when the specific con-ditions pertain but not every year. The farm situation is so fluid and uncertain that simplistic logic can never be enough. The famous unpredicted "Black Swan" event may not have been included in the mechanistic model.

This would open the way for artificial intelligence or machine learning to take over and work on a best outcomes model, but this could lead to a different type of failure if the training set did not cover uncertainty.

Risks in machine learning or artificial intelligence

Digital agri-technology is a broad term covering a huge range of systems that control machines that analyse internet data and replace human drudgery in repetitive clerical tasks. They all use software. As systems have become more complex, machine learning software has become a routine tool for optimising performance and reducing the options presented to an operator. These systems compute probabilities of associations and use patterns to identify the most probable cause. They are not able to deduce causality directly. In other words, they cannot look outside the box. The box being the set of input data they are given.

Akter et al. (2021), analysed the risks of bias in AI systems particularly in human management systems, analysing credit worthiness, surge pricing at Uber and candi-date selection. They were most focused on bias against humans in different ethnic or social groups; however, this is just a subset of incorrect decision making and we can use the same methodology to look at possible risks. They identify three main sources of risk:

- Training data bias
  - too narrow a set of parameters, too few data points

- Method bias
  - the correlation fallacy identifying the wrong cause of an input
  - confirmation bias
  - automation bias
- Societal bias
  - stereotyping, implicit associations, prejudice

In agri-tech systems, the first two are most important although buying software developed in territories with different attitudes to animal welfare or crop acceptability will have an impact on the decisions a system will make.

At present, these are theoretical risks, but it would be wise to ask the question: what was the nature of the data set used to train the system? If, for example, an animal recognition system is trained on one breed, how reliable will it be on other breeds? This type of error has typified facial recognition software where the training sets were the white males in the development lab. Will the weeds be recognised correctly by an autoweeder if the training images are captured in different light conditions?

Method bias is harder to identify but, for example, an automated combine harvester will need many sensors to detect correct performance and these will have to operate correctly in the dusty conditions of harvest. There also may be unexpected content in the harvest — animals, stray humans, etc. The software would need to be resilient and not just stop the machine to call a human who may not be immediately available to deal with the anomaly. Digital Methods will need to be very reliable and the feedback well understood if a problem is not to be repeated.

Software incompatibility

Digital technology is very often developed by start-up companies in a surge of enthusiastic invention. Very often the first stages are developed in isolation to test an idea to see if it works. This stage of coding is not likely to have not been designed to link to other systems, which increases the risk of incompatibility. Modern software is designed to be modular with application protocol interfaces (APIs) for the different layers of interconnection. Good practice is for these to be published and made available to the developers of potential connected systems. In tractor-based systems, all major manufacturers conform and contribute to the ISOBUS standards, but many of the new start-ups do not appear to do so, seeing their systems as standalone. In the dairy sector, many manufacturers and software developers are hesitant to enable their systems to connect, preferring to be monopoly providers to the customer. This then leads to incompatibility with the farmers' systems for veterinary, crop and business management.

Limited interface specifications for new developers

Digital systems rely a great deal on system modules that have been developed by others, for example, the operating system for the microprocessors; the pace of development has been so rapid that legacy software issues have been a constant source of worry. In my development of rumen telemetry, I relied on using Android phones as a platform to connect to the radio via a USB plugin using the On The Go (OTG) protocol. However, as the smartphone became more aimed at being an entertainment toy for consumers, the software support for OTG through the USB port was limited or non-existent. The Android OS was changing every year and in the end in 2020 I had to re-design the system to use a more stable LoRa-based system.

The modular APIs and standards for new technology are not settled and stable. Companies that do not create open interfaces either have to dominate a sector (for example, the way that Apple products do) or use open standards such as ISOBUS that encourage specialisation and modularity.

Digital control in animal management

The ethical risks of damaging a plant or soil system are limited in ethical impact in comparison to the risks inherent in animal agriculture where the health and welfare of sentient beings that feel pain are the issue. At present, legislation requires that animals are inspected daily by a human herdsperson, and although this may be often ignored, it does legally prevent any system being introduced that aimed to replace the human with an automated system without a change in the law. That is a legislative risk and would have to be fixed as, for example, the requirements for animal handling were changed with the introduction of robotic milking.

Environmental control

Many animals, particularly poultry, are kept in artificial environments. The parameters for animal comfort are controlled electronically with a control system designed to keep some comfort parameters (temperature, humidity, etc.) within a defined range using fans, vents to change air flows and simple sensors to measure the comfort parameters.

The control algorithms use models of the comfort levels of the animals for optimal comfort and thus production. The models are largely based on empirical experiments conducted during the second half of the 20th century. These models may be updated as new breeds and strains emerge. The risk here is that we have two evolutionary processes running against each other. The animals may be evolving traits to survive in the environmental conditions of intensive agriculture even as they are being bred for efficient feed conversion. Simultaneously, the control systems of the buildings may be trying to optimise production. We may inadvertently cause an arms race between two systems one of which is made up of sentient animals and the other is a machine intelligence trying to optimise production.

Do we really want to be breeding sentient animals to survive conditions which would be highly unnatural to their original farm ancestors in the pursuit of cheap protein for humans? The pejorative term factory farming has already led to legislative interventions which may not have improved animal welfare but seem better to the public and politicians. The risk here is of an ethical dilemma.

Animal welfare monitoring

Animal welfare is a major concern for the food retailers and they are increasingly turning to sensor data to detect problems on farms. For example, lameness in dairy cows is currently checked as a percentage determined as lame on a periodic inspection visit. This could be replaced by automated systems. However, systems can make mistakes, sensors become blocked and software can 'model' what it thinks is happening. Since the meeting of the supply chain criteria for good welfare such as a low percentage of lame cows or no overcrowding of poultry is a contractual obligation, which if the farm fails, it can face ruin and bankruptcy. There is a risk that sensor or software errors reading them could determine the future of the farm.

Control systems have been developed to maintain the environmental conditions within intensive animal husbandry buildings and these have usually been based on Programmable Logic Controllers which have limited capabilities for reading sensors and using proportional integrated and derivative (PID) control algorithms to change ventilation parameters. These systems were originally free-standing and normally overseen by a human operator periodically who can apply common sense to recognise when things are not correctly set. More modern systems are internet linked and parameters can be viewed remotely. Feeding systems can also be automatic. There will almost certainly be attempts to apply machine learning or in marketing terms artificial intelligence (AI) to these systems. AI systems are supposed to learn by feedback results and optimise conditions to achieve desirable outcomes. If the desirable outcomes are not set with animal welfare considerations, then there are major risks that the system will be driven to optimise output. At present we only have proxy measurements of animal welfare — such as thermal comfort. As systems become more complex, it is inevitable that there will be occasional failures and it will be hard to show what the cause of the problem was.

It is not always possible to know the outcomes of bad software code in advance — test structures do not tend to look beyond ensuring that the code is sound but not what its effects will be when inputs and outputs do not conform to predictions.

Thus, there are risks in systems heavily dependent on software which need the ability to operate safely.

Generics
Data security risks

I was at a farm with robotic milking doing some sensor installation work and was asked by the herd manager if I had moved a cow from one group to another. This could be done by software and would require a login. I was of course innocent of making any changes and by analysing the network log it was possible to show the herd manager that these changes had been done from an offsite computer using a staff login. The change could have been done accidently by a child or a tinkerer. The conclusion was that a former member of staff still had access to settable parameters on the robotic milking allocation system. Thus, data security can have an instant impact on animal welfare.

A security breach by a malign third party that corrupted the data could put a farm business at great risk, quite apart from a huge loss of privacy. Not only would confidential data about the farm business be visible, but there is the potential that feed parameters could be changed, animal treatments hidden and major damage done, for example, by contaminating the milk with antibiotics from treated cows by switching off the flags for contamination.

Cyber attacks on infrastructure are now a routine and as always the best defence is disciplined password security and physical security of the systems. It is no longer science fiction to suggest that state actors in hostile countries can interfere in food production without leaving home.

Product liability

When a machine is controlled by a human the liability for an accident or loss generally starts with the driver. The owner may have to prove that the machine is faulty as a result of the manufacturer's negligence and be able to pass the liability to him. However, if the accident is caused by software developed by another party using sensors provided by another supplier, then the liability disputes could become very complex. If an autonomous vehicle causes a loss, it is inevitable that litigation will follow. Until the liability issues are settled, it will be hard for widespread adoption of new technology to proceed with confidence.

Legislative risk

In commercial software development, there are no constraints by regulation on what software can or cannot do. Features such as limit controls would be set by best practice but there is no regulation — for example, rumen telemetry boluses had no governing standards before they were invented.

There is a legislative gap caused by the pace of invention. I was told in 2007 that a telemetry bolus that I had invented and wanted to sell was not a Veterinary medicine and was not covered by the Veterinary Medicines Directives. Neither was such a device covered by medical device legislation. So we launched it anyway; it was an unregulated market.

Position navigation timing failure

The first Global Positioning System (GPS) system was established by the US military in the 1980s and others have followed suit using a similar design. A fleet of satellites is deployed around the planet each with a highly accurate atomic clock constantly sending a radio signal of the time and position. A ground-based receiver identifies at least four satellites and compares the time from its own clock to that from the satellites and thus by vector trigonometry can determine its location to within 7 m on 95% of the Earth (FAA, 2021). The accuracy can be enhanced with more satellites in view and by fixed ground-based stations to provide a corrective signal in local areas. Other systems have been put into position giving either global coverage Galileo (EU), Beidou (China), GLONASS (Russia) or systems giving local coverage over Japan and India.

Precision location for field operations, vehicle navigation, virtual fencing and even livestock location thus requires a network of communications of very weak radio signals that could easily be disrupted by a number of factors.

There is interference and attenuation due to changing atmospheric conditions although this can be predicted and compensated for by suitable metadata and algorithmic correction.

One natural risk is that solar flare activity can either destroy electronics in space or even receivers on the ground. There is considerable interest in these events[1] and meteorological offices now provide space weather forecasts (Met Office, 2021; NASA, 2021) which will give a few hours warning of major disruption. As the worst regular disruption occurs over the polar areas, it does not affect agriculture seriously although the loss of satellites would be a major problem which would be fixed by having redundancy and spare satellites ready to be deployed. With new launch facilities being built and satellites getting smaller and cheaper, this is becoming less of a problem.

The biggest potential disruption will come from a political or military crisis that could cause the systems to be encoded to prevent civilian or hostile powers using the signals or even for satellites to be attacked. The major powers now have military space capabilities which are a potential disruption.

For this reason, countries such as the United Kingdom are developing national capabilities to provide local PNT and signals security (UK Government, 2021). It seems likely that other countries will also for reasons of national security set up similar localised systems. This could cause the system developers problems if each different system operated with different encoding. In a perfect world, the receiver and software would be compatible across all systems. There is a risk of failures of interoperability.

Labour availability risk

One of the key drivers of developments is the shortage and cost of labour. In the United Kingdom, employment contracts in dairying (which is well paid) have been available in Polish for 20 years indicating a major source of extra territorial

---

[1] https://www.spaceweatherlive.com/en/solar-activity/top-50-solar-flares.html.

recruitment. For example, the chief recruiting areas for seasonal workers in fruit and vegetable production in Europe have been the former countries of the Soviet Bloc in Northern Europe or from Africa in the South. A similar pattern of migration from countries south of the Rio Grande pertains in North America.

However, political resistance to immigration and movements such as Brexit have seriously disrupted these flows in the United Kingdom and the United States of America.

Politics and economics change over time and if the availability of labour rose then some robotic technologies would cease to have a cost benefit. This would go against all long-term trends but it could still happen.

Skilled labour shortage

Although digital systems replace labour in the fields, they demand highly skilled service technicians and installers to be available on call within the countryside. This poses a training and recruitment problem as the farming industry is widely perceived as primitive, backward by the type of technician working in computers and robotics. Even the existing machinery dealerships tend to have mechanical engineering skills rather than electronics and computer skills. Shortage of skilled labour could slow down the uptake and the quality of support available to the new technology.

The management of farms in developed countries tends to be dominated by elderly owner managers who are very often resistant to taking up new technology and have no technical interest in computers. This could be a major barrier to adoption. Craft level training in electronics, sensing and computers is not consistently provided by the traditional agricultural colleges. Although young people are heavy users of smartphones, they do not have much understanding of their capabilities and so are thus helpless in the event of software failures.

Rural internet bandwidth and speed

Digital agri-technology, particularly systems needing cloud computing, needs fast internet connections, and unfortunately even in densely populated industrialised countries like the United Kingdom, rural broadband coverage is incomplete. Suppliers focus on areas with a high density of high frequency users streaming video both for fibre optic and airband services. Often the last kilometres are left to legacy copper telephone wires which have inherently lower speed than fibre optic.

Alienation and the death of farming culture

Human culture around the world has long been interwoven with the work associated with growing and harvesting food. Literature and art are full of imagery of sowing and reaping of grapes and sheaves, cows, sheep, eggs and lambs. Many of the parables of the Christian gospels rely on analogies and references to the work of farming and the rhythm of the seasons.

For many years, the disconnection between what people believe about how food is produced and the reality has caused friction and alienation between producers and consumers. This can surely only get worse as humans disappear from the country-side to be replaced by robots. The political heft of rural voters has diminished as well and politics is often driven by pressure groups with agendas that have little rela-tionship to practicality, being about how people feel rather than what is real.

Humanity may lose all contact with the natural world except through farm park-lands preserved as a bucolic ideal with food production being done in specialised areas and under cover to control the growing crops or animal. This could have a serious impact on the human psyche with unpredictable results, for example, with increased mental illness or the promotion of disruptive food fads.

Discussion and conclusions

The assumption that digital agri-technology will be wholly beneficial and relatively risk free is probably naive. Many assumptions about the future impact of new tech-nology have proved false in the past. No one predicted that the continuous availabil-ity of mobile data would lead to less understanding of what was happening in the world, or that algorithms to give users more of what they liked could lead to expert advice on vaccines being ignored or that governments would sponsor misinforma-tion sites. We can only predict developments on what we know currently works and a few extensions of it.

I this Chapter, I have tried to break down the risks into categories to simplify the description, but many of the risks overlap each other. As a techno-optimist, I am convinced that digitisation of systems in agriculture will fix some of the problems created by the chemical and biological innovations of the 20th century, but it could be a bumpy ride.

References

Akter, S., McCarthy, G., Sajib, S., Michael, K., Dwivedi, Y. K., D'Ambra, J., & Shen, K. N. (2021). Algorithmic bias in data-driven innovation in the age of AI. *International Journal of Information Management*. ISSN: 0268-4012, *2021*, 102387. https://doi.org/10.1016/j.ijinfomgt.2021.102387

Crookes. (1898). *The wheat problem* (authors collection).

FAA. (2021). https://www.faa.gov/about/office_org/headquarters_offices/ato/service_units/techops/navservices/gnss/gps/.

Girdžiūtė, L. (2012). Risks in agriculture and opportunities of their integrated evaluation. *Procedia - Social and Behavioral Sciences, 62*(2012), 783−790. https://doi.org/10.1016/j.sbspro.2012.09.132, 1877-0428.

HSE. (2021). https://www.hse.gov.uk/agriculture/pdf/agriculture-fatal-injuries-2021-summary.pdf.

Lee, D., & Hess, D. J. (2020). Regulations for on-road testing of connected and automated vehicles: Assessing the potential for global safety harmonization. *Transportation Research A: Policy and Practice*. ISSN: 0965-8564, *136*, 85—98. https://doi.org/10.1016/j.tra.2020.03.026

Met Office. (2021). https://www.metoffice.gov.uk/weather/specialist-forecasts/space-weather.

Mummery, C. L., van de Stolpe, A., Bernard, R., & Clevers, H. (2021). Chapter 6 - Cloning: History and current applications. In C. L. Mummery, A. van de Stolpe, R. Bernard, & H. Clevers (Eds.), *Stem Cells* (3rd ed.). Academic Press, ISBN 9780128203378. https://doi.org/10.1016/B978-0-12-820337-8.00006-X

NASA. (2021). https://www.nasa.gov/mission_pages/sunearth/spaceweather/index.html.

UK Government. (2021). https://www.gov.uk/guidance/space-based-pnt-programme.

Index

Printed in the United States
by Baker & Taylor Publisher Services